WE ARE MOVEMENT
WE ARE MOVEMENT
WE ARE MOVEMENT
WE ARE MOVEMENT
WE ARE MOVEMENT
WE ARE MOVEMENT
WE ARE MOVEMENT
WE ARE MOVEMENT
WE ARE MOVEMENT
WE ARE MOVEMENT
WE ARE MOVEMENT
WE ARE MOVEMENT
WE ARE MOVEMENT
WE ARE MOVEMENT
WE ARE MOVEMENT
WE ARE MOVEMENT

AF060017

Praise for *We Are Movement*

'Wayne McGregor is not only a world-renowned choreographer, but a deep thinker about how the body shapes the mind with a unique ability to link his creative art with science and philosophy. His dazzling fluency across disciplines is on full display in his remarkable new book. *We are Movement* is not only an innovative meditation on the embodied, embedded mind, but a practical guide to thinking, and living, better. Best read while stretching' Anil Seth

'This a deeply humane book about how we move and touch the world and I loved it' Edmund de Waal

'Everything about us, including our intelligence, flows from and through our bodies. In this era of AI, Wayne McGregor offers a powerful reminder of the original general purpose technology: the human body itself. Our movement, he argues, matters. But this "physical intelligence" isn't just about motion; it's a complex, cognitive system that underpins our creativity and connection. *We Are Movement* is a compelling manifesto that reveals why our most powerful software remains written in flesh and bone' Mustafa Suleyman

'This book invites us to be alive, alert and aware; to be grounded in the body and its innate intelligence – a gift to our urban animal natures' Antony Gormley

'Wayne McGregor's *We Are Movement* celebrates the profound and often overlooked concept of physical intelligence, arguing that our bodies hold a deep, intrinsic knowledge. In this book, McGregor highlights how mastering body language, and intentionally holding the space around us, can unlock extraordinary new levels of creativity and self-understanding, fundamentally revolutionising the way we live and communicate. This is an insightful book for anyone seeking to experience life with greater fluency and presence. I strongly recommend it' Carlos Acosta

'Movement is fundamental to living life well, with passion, strength and happiness. As Wayne explains in his superb new book, physical intelligence allows us to make meaning of our everyday' Darcey Bussell

'In *We Are Movement*, Wayne McGregor distils a lifetime's enquiry into the potential of our physical being into a readable overview, presenting many fascinating ideas along the way. The book is full of refreshing and valuable insights, and reads like an advanced user's manual for the human body' Max Richter

'In Wayne McGregor's new book *We Are Movement* he brings together with such skill the connection between brain and movement and the relationship between body and mind. The chapter entitled "We Are Balance" which helps understand the science on how to fall particularly resonated with me because falling in old age is such a universal problem. This will empower me to tap into all of the movement skills I developed as a dancer, to live a longer, stronger life. I first saw Wayne's choreography in the early 90's with his extraordinary dance company Random, it blew my mind. I knew I was looking at the future of dance, and I was right. This book is a revelation, and a must read for developing our ability to move, unlocking our physical intelligence and the confidence that follows' Arlene Phillips

'Wayne McGregor's whole life is devoted to analysing how movement can be controlled to create beautiful dancing shapes. From his unique perspective, he is now helping us non-dancers discover ourselves ... now McGregor challenges us to change our lives for the better' Sir Ian McKellen

'When I work with Wayne I gain a deeper understanding, not only of character, but of who I am, and who I can be as a performer, an artist, a human. He has shaped how I work in such a fundamental way. I am in awe of this man and his brain' Saoirse Ronan

'Despite never having become a dancer professionally yet completing training in my early twenties, the self-awareness I have of my body has helped and informed everything I do from work, to walking, to lifting, to childbirth. Physical awareness is important to all of us to be healthy, fully functioning humans whether we sit behind a desk all day or rotate from the ceilings on a rope! This is a revolutionary book which will improve the lives of all its readers ... highly recommend' Paloma Faith

'Another tour de force from Wayne McGregor; this time it's prose that moves. *We Are Movement* diagnoses and treats modern ennui by skilfully empowering us to reflect on what makes us human. It's a masterful masterclass: every page is by turns prescient and practical. Wonderful!' Aleks Krotoski

WE ARE MOVEMENT

WE ARE MOVEMENT
Unlocking Your Physical Intelligence

Wayne McGregor

TONIC
LONDON • OXFORD • NEW YORK • NEW DELHI • SYDNEY

BLOOMSBURY TONIC
Bloomsbury Publishing Plc
50 Bedford Square, London, WC1B 3DP, UK
Bloomsbury Publishing Ireland Limited,
29 Earlsfort Terrace, Dublin 2, D02 AY28, Ireland

BLOOMSBURY, BLOOMSBURY TONIC and the Tonic logo
are trademarks of Bloomsbury Publishing Plc

First published in Great Britain 2026

Copyright © Wayne McGregor, 2026

Wayne McGregor is identified as the author of this work in
accordance with the Copyright, Designs and Patents Act 1988

All rights reserved. No part of this publication may be: i) reproduced or transmitted in any form, electronic or mechanical, including photocopying, recording or by means of any information storage or retrieval system without prior permission in writing from the publishers; or ii) used or reproduced in any way for the training, development or operation of artificial intelligence (AI) technologies, including generative AI technologies. The rights holders expressly reserve this publication from the text and data mining exception as per Article 4(3) of the Digital Single Market Directive (EU) 2019/790

Bloomsbury Publishing Plc does not have any control over, or responsibility for, any third-party websites referred to in this book. All internet addresses given in this book were correct at the time of going to press. The author and publisher regret any inconvenience caused if addresses have changed or sites have ceased to exist, but can accept no responsibility for any such changes

A catalogue record for this book is available from the British Library

ISBN: HB: 978-1-5266-2953-1; TPB: 978-1-5266-2954-8;
EBOOK: 978-1-5266-2952-4

2 4 6 8 10 9 7 5 3 1

Typeset by Six Red Marbles India
Printed and bound in Great Britain by Clays Ltd, Elcograf S.p.A

To find out more about our authors and books visit www.bloomsbury.com
and sign up for our newsletters
For product-safety-related questions contact productsafety@bloomsbury.com

Without you, none of this would be. Ella, Lawrence, Antoine, Freud, Mies, Orez and Zero – a family like no other.

Contents

Introduction   xiii

PART ONE: MOVEMENT is AWARENESS

1 We are Attention   3
2 We are Unique   20
3 We are 3D   38
4 We are Balance   56
5 We are Co-ordination   69
6 We are Chemistry   84

PART TWO: MOVEMENT is COMMUNICATION

7   We are Touch   115
8   We are Presence   138
9   We are Gesture   153
10 We are Stories   173
11 We are Memory   195
12 We are Empathy   214

PART THREE: MOVEMENT is CREATIVITY

13 We are Fearless   233
14 We are Invention   253
15 We are Collaboration   269
16 We are Fluency   278

Acknowledgements	289
Notes	293
Bibliography	297
Index	301

Introduction

For millennia, we have undervalued the wisdom of our bodies, but as the last century dwindled, researchers spurred by the advance of technology came to recognise the staggering breadth of the body's knowledge. Cognitive scientists, psychologists, philosophers, and linguists finally dismissed an idea that was once considered gospel: that the mind and body are separate entities. Today, it is widely accepted that our thinking is physical: it occurs through and with the body.

In some fundamental way, I have always known this. After all, I've had dance in my life since childhood. I have run, jumped, flowed, and otherwise used my body beyond the prosaic since infancy. Free to move, free to be. I don't recall my parents ever telling me to sit still. It was the opposite – they would constantly encourage me to explore any physical activity: cycling, swimming, gymnastics, long-distance running, tennis, badminton, squash, aikido, skateboarding, the list goes on.

'Try it and see if you like it,' my mother would say, with only one condition – I had to commit to the activity for a decent amount of time. Much like her strategy for getting me to eat broccoli, which involved trying it five times to see if I would develop a taste for it. I still don't love broccoli, but I did develop a taste for almost every physical activity I attempted. Consequently, my parents spent my childhood ferrying me from club to group to class – a dedicated physical endeavour of their own.

I recently discovered an old photograph of myself in my country-dance team at school. I must be about seven years old in it. I had long thought that I was inspired to take up dance by the physical fluency of John Travolta's performances in *Grease* and *Saturday Night Fever*, but looking at the photo I realise I have been misremembering the beginnings of my dance dream. John Travolta had come later. It was actually when I was in primary school that the passion was ignited, dressed in shorts, a white shirt, a blue tie, and black plimsolls, learning those English country dances around the maypole. Not basic dances but intricate group formations requiring me to touch, cooperate, and work with other bodies in synchrony to lively taped music with an easy-to-follow beat.

I smile as I look at the photograph, recalling the sheer joy of those lessons! I looked forward to them and took them seriously. While other classmates were messing about and refusing to dance with the girls, I was laser-focused and often frustrated that the teachers weren't teaching us the new routines quickly enough. I've always been impatient.

We shared these dances at school shows but also toured them. Those town halls packed with appreciative audiences of mums and dads clapping and cheering felt like stadiums.

INTRODUCTION

It was at these events that I really got the dancing bug – right there in the grey Northern towns of the late 1970s.

On reflection, country dancing gave me far more than the thrill of performing. It is a highly organised dance form. It demands precision, spatial accuracy and split-second timing. You have to find a collective rhythm to the music and stay in time together while executing a range of steps and movement combinations. Your hand-eye coordination, for weaving the maypole ribbons and finding each partner's hand as you dance past exchanging places, needs to be extremely fast. Yet the facial expression you need to present when performing has to be relaxed, calm and effortless – never a gnarled grimace of fearful concentration. These dances take practice to accomplish, and, rewardingly, they improve over time with hundreds of repetitions. They're also a great example of distributed cognition – an interconnected kinetic human organism that works and flows as one to create a beautiful effect. In country dancing, I was physically thinking and thinking physically. I wasn't aware of it at the time.

Physical thinking – thinking through and with the body – is any reasoning that stretches beyond the central nervous system alone. Another way to conceive of physical thinking is as something that occurs whenever the body helps the brain out. Or as the body being 'thoughtful' and using its intelligence to direct motion in concert with the brain. Physical intelligence refers to the potential of our body and mind, while physical thinking examines how we effectively use that potential. Both serve as the focal points of this book.

For all the advancements of recent years, we still comprehend so little about how the human body and mind function. For most of us, most of the time, they simply do. We accept the magic and mystery of the body's technology without thinking about it. It is only when something goes awry – when we break an arm, cut ourselves, or suffer from an inexplicable tremor – that we start to enquire further. For most of us, an 'out of sight, out of mind' approach is the default. It is as if we fear the invisible, the unknown, the alchemical forces inside our own bodies, instruments that we have not taken the time to understand.

And yet, our subjective way of being in the world – experiencing it, engaging with it, sensing and perceiving it – relies on this sophisticated and innate intelligence: our physical intelligence. This physical intelligence is expressed in any form of behaviour, information, process, and reaction that the body *knows,* and it is instinctive, pre-verbal, and continually upgrading itself. It's a form of intelligence shared by us all; it manifests far beyond the dance studio and, in conjunction with linguistic and logical intelligence, enables us to synthesise our mind and body experiences into diverse embodied interactions – what we more usually call life.

By studying and practising this intelligence, beginning to uncover its potential, and (re)discovering its magic, we can release the knots in our emotional and physical selves, stimulate mental and tacit flow, and ultimately free ourselves to experience new forms of creativity and connection.

After decades of exploring the dimensions of our physical intelligence through physical thinking in the studio

INTRODUCTION

with professionals and amateurs of all kinds – dancers, actors, singers, children, older individuals – and increasingly also with scientists, anthropologists, geneticists, technologists, and designers, I have become expert in reading and interpreting movement, picking up on and analysing the non-verbal and harnessing it to explore individuals' interior landscapes, capturing their unique physical signatures and, in turn, collaborating with them to use their bodies to problem-solve and innovate.

In this book, I aim to share my fascination with how bodies think, to intrigue you with how people move and, by extension, how we interrelate. This book is an invitation for you to enter the realm of your own physical intelligence and utilise that understanding to enhance your physical thinking skills and proficiency.

In Part One: Movement is AWARENESS, all readers – regardless of age, ability, or fitness level – are called to become more attuned to their physical sensations and the physical intelligence behind everyday actions. As we increasingly direct our awareness towards screens, our consciousness rarely checks in with our flesh. Part One will entice readers to be more present in their bodies while drawing awareness to the rudiments of physical thinking. It will also explain precisely how these mind–body interactions occur using scientific theory and introduce readers to the practice of manipulating attention through physical thinking.

Having peered deep within, we turn our gaze outwards in Part Two: Movement is COMMUNICATION. This section of the book will shed light on how physical thinking affects our relationships, placing readers firmly in dialogue with the world. We cannot communicate without movement; everything we say comes as a result of physical thinking. As well as unpacking the processes that allow us to convey thought and emotion through gestures and physical contact, this section will interrogate the ways in which our bodies control how we are perceived by others, at times, irrespective of what we intend to express.

Part Three: Movement is CREATIVITY will zoom in on the nature of creativity as a skill that can be developed by absolutely anyone, with applications in every thought process and activity. Thinking through and with the body is crucial to its advancement. I will gather the most illuminating and helpful lessons from each prior section and consider how they relate to one another to illustrate the astonishing richness of a physically fluent existence.

Throughout, I will draw on learning from the fields of dance, cognitive and social science, performing arts, medicine, technology, entertainment and sports to investigate the body's knowledge and show how much physical thinking we already do, from the elementary to the virtuosic. We will see how our bodies communicate through unique physical signatures and cognitive systems and discover the reasoning behind socialised inhibitions and fears that prevent us from living fully, physically. We'll learn about tools for creative thinking that expand imagination and help us to overcome ineffective habits and harmful behaviours and seek out cutting-edge technologies that promise to amplify our natural aptitude. We will come to understand how living

INTRODUCTION

a physically fluent existence will benefit and empower each and every body.

★ ★ ★

As the contract for this book was being signed, I turned fifty. My birthday came at the beginning of the first UK coronavirus lockdown, so all my celebrations were cancelled or stalled. In a flash, all parties, plans, shows, events, sessions, lessons, and meetings – all of our movements – were restricted. We were shut in and ordered to stay at home. Total stasis.

Like most, I started to look inward. Smaller. My whole body felt miniaturised like I had been put in a sci-fi shrinking device. It was an intolerable feeling. I was bound. This was a first for me.

Many people find themselves bound – some, like me, only during the initial Covid lockdown or periods of significant stress, others for an entire lifetime. No appetite for the unfamiliar. No curiosity or passion for change. No energy for movement – either in body or in mind. So, much as after successive lockdowns we finally shook the restrictions and habits we no longer needed, we must now recognise the limits we place on ourselves through fear. We can free ourselves through movement. This is not only done literally by moving, although this is a great start, but by embracing movement and our physical intelligence in its widest sense.

We Are Movement encourages this. We can cultivate our attention so that we are not only more focused but also better attuned to our surrounding environments and internal sensations. Through more open, honest, and authentic physical communication, we can enhance our capacity to be good friends, partners, parents, professionals, and members of society. Creatively, we can mobilise our senses, imagination, and bodies for fresh and original purposes. We can harness our ability to access our physical fluency at will, master spontaneity, and navigate daily life fearlessly. To dance freely. To live fully.

This freedom is ever more critical in our brave new world of AI, bodily movement tracking apps, wearables, robotics, biosensors and shape-shifting interfaces. Here, the currency of physical intelligence is prized, and rightly so. And here lies the greatest potential for this book: to be a user-friendly guide to the body and the potential it holds. I would be delighted to be your physical intelligence mentor, to pilot you through this thrilling form of multi-dimensional thinking with and through your body, and in doing so, encourage you to be your most unlimited self – to unveil the true dancer within.

PART ONE

Movement is Awareness

I

We are Attention

For the second-year secondment programme at my university, we had the opportunity to select a company, theatre, or arts organisation where we would spend six weeks in residence. I wanted an experience that balanced the artistic and the managerial: I wanted to be close to dancers and choreographers, to be dancing, and to learn about the broader mechanics of running a creative business. Luckily, I was invited to the US to spend time with Dance Alloy and the Pittsburgh Dance Council.

After an incredible first week there, I felt energised and alive. And then I dislocated my knee. I wish I had a spectacular story from the studio or stage about a complex and challenging movement that caused the injury. But alas, I tripped on the kerb. As I fell, my leg folded peculiarly: my kneecap migrated southwest, and my leg bones northeast. I could feel the end of my tibia sticking out through my skin. Crumpled by the side of the pavement, I was in agony. Confused, disoriented and crying. What happened next was a blur. There was waiting, endless waiting on the sidewalk, then an ambulance, hospital, terrible language from me, pain medication, X-rays, scans, prodding, poking and finally, a doctor who cupped the offending patella and slid it back into place while easing my heel skywards. Relief. Later, the knee was immobilised in a rigid metallic leg brace, which I was told I must keep on for six weeks. That done, I was sent home on crutches. What a secondment.

My rehab took considerably longer than I expected, not least because during the knee immobilisation, the muscle tone in my legs wasted away to a limp shadow of itself (this was the 1980s, when treatment was… different). Although my upper body strength changed for the better, my leg muscles weakened, and without the crutches, walking was unstable. But the injury and the three-month rehabilitation taught me so much about paying attention to my body. There were so many new sensations: my first 'out of body' feeling while adjusting to a new lack of mobility, the incessant itch and desire to scratch under the knee brace, the relief when I removed the brace to shower, and the feeling of awkward action when learning to use the crutches. I had to redistribute my weight, not just physically but psychologically, literally shifting everything away from the injured side, and I was forced to practise unusual co-ordinations. It was fascinating how quickly my brain and body adapted, finding workarounds for my new normal. My speed of action steadily increased, my coordination settled, and my balance became so attuned I could use the crutches as startling new pivot points for choreography. Overall, and surprisingly fast, I found a new physical economy and, indeed, new degrees of freedom. Something lost, something gained.

This brutal reminder of the inherent fragility of the human body and mind, along with the nurturing they both require, marked a pivotal moment for me. Upon returning from the US, I was determined to incorporate my experiences into my daily practice in the UK. From then on, I adopted a research-based approach that integrated technical and choreographic training with the latest scientific and psychological insights, inviting a new cross-disciplinary method into my projects. The exposure to this nexus of brilliant minds working outside of dance but with immense relevance to it has fuelled my commitment to understanding and developing my own physical intelligence and advocating for the evolution of everyone else's.

How physical damage can occur from a simple and momentary lack of attention continues to interest me. Ironically, it was a moment of inattention, of not being present in my own body, of not being present in my surroundings, that marked the beginning of my journey into mining what the body knows.

Attending to Attention

It is in a dancer's interest to pay close attention to their body. I learned that in Pittsburgh. We need to be in constant, close dialogue with every aspect of our physical intelligence, noticing and updating what our bodies tell us. By doing this, we can learn complex choreography, enter the right frame of mind to perform at the highest level, and maintain our health and well-being.

Dancers know that the body is an astonishing instrument that collects and collates feedback all the time. We know that if we tap into that and start to understand some of the feedback mechanisms in a more detailed way, then we do not just expand our physical vocabularies; we give ourselves a chance to experience benefits to our health and well-being, the beginnings of what might be called a dictionary of physical intelligence. Integrating multiple aspects of physical intelligence also allows us to dance creatively, make meaning in and through our bodies, and share and communicate our inner selves and emotional truths.

And here's the thing: this kind of attention is not the preserve of dancers alone. Every single second of every single minute of every single day of our lives, we cycle between multiple modes of attention to our bodies. We wake and search for the floor with our feet; we squeeze the toothpaste along the slim line of our brushes and scan the congested street before running for the bus. Over time, we become experts in some forms of physical awareness that align with our lifestyle and human habits: climbing the stairs, opening doors, sitting, standing, gaming, golfing, page-turning, thumb-sucking, fingernail-biting and, yes, breathing. But there are many others we neglect. When was the last time, for example, you thought about backspace – the volumes behind your body?

This neglect is a shame, partly because it cheats us of experiential richness but also because it limits our physical potential, increases our chance of falling ill (or tripping over a kerb) and leads to our embedding subconscious (mal-)adaptations into our physical intelligence that then become a 'new normal'. And yet, it doesn't have to be this way.

WE ARE ATTENTION

I want to help you become more aware of your physical sensations and more present in your body. What we attend to frames and colours our existence. Without giving something conscious attention, inviting it into awareness, or simply noticing it, it eludes us. I also want to help you capitalise on your unique physical signature to access deeper layers of experience. This is work that anybody can do regardless of age, ability, or fitness level.

Paying a new kind of attention to your body can be the beginning of a different relationship with it. It starts a journey towards a more complete embodiment, an embedded, unconscious understanding, felt rather than thought. It means experiencing the body in its totality, as an interconnected whole, rather than a kind of archipelago of isolated parts.

Perhaps you feel that you are already attending to your body: you follow an exercise programme, you run, you resistance train, you do yoga, you CrossFit, you swim, you practise tai chi; you have a motion data collection device that records your steps, your calorie burn, your heart rate, your levels of insulin, your oxygen states. In short, you have sought liberty through physicality, the unceasing motion within, of, and by the body whereby you feel alive, energised and well. My eighty-year-old parents' quality of life has changed dramatically since they bought a Fitbit. They diligently work towards their 10,000 daily steps, vastly improving their mobility and aerobic fitness. However, this kind of data-driven attention to *doing* is no replacement for attention *sensing*. My parents' quantitative obsession with their step count has compromised their qualitative abilities to access their physical intelligence more holistically. I have seen this compromise in many circumstances: where the yoga teacher focuses on the stretched hamstring rather than the breathwork, where the gymgoer is encouraged to lift heavier and sacrifices the resistance of the return motion, and where the runner heavily pounds the streets, speeding towards the finishing line and a case of shin splints without ever tracking the force of their foot on the pavement.

I have also witnessed it in the dance studio – the obsession with a certain kind of virtuosity articulated in more turns, higher jumps, flatter pliés and over-extended splits. A drive towards the quantitative and away from the qualitative. This attitude toward our bodies needs to be upended. And this requires a mind shift where the *what* (quantitative) and the *how* (qualitative) are brought into balance. Where the *how* becomes the new gold standard.

The *how* is what I focused on while on crutches. I thought less about mobility mechanics and more about a gentler grip force. The least amount of effort for the greatest gain, resulting in fewer hand callouses. I tried to pay qualitative attention when swimming; instead of mentally tallying my laps, I endeavoured to feel the resistance of the water and relished the variation in the speed of my stroke.

Attention, then, is both a mental and physical faculty, a means of accessing and engaging with our external environment and embodied selves. In this section, we'll zoom in on key aspects of these processes as they play out in and through our bodies and minds. How we attend to the world – outside and within – fundamentally steers our behaviour by framing and tinting our every thought, feeling, and action, illuminating our present and future potentials.

Dancers often start their day by lying down, mentally checking in with their 'today body' in a head-to-toe sensation audit before their obligatory physical technique class. This visualisation ritual is a kind of internal body reset, an attentional breath before the exertions of the high-performance day ahead.

I walk most early mornings with the dogs – my reset. That is the time when I can make my own sensory check-in. We have two inquisitive whippets who love to be outside and resist routine in their walking routes. They like variety – city, beach, and countryside – where their noses are activated, and their ears are alive. Our whippets have exceptional situational awareness.

Often, when I walk alone, I deliberately take unfamiliar routes. My phone is in my pocket, not in my hand, and I do not walk and talk. I try not to be on autopilot. I try to focus on things other than the tasks ahead or my next meeting. I do not meditate as such, but my walks exist in a similar sphere. I walk slower than usual, with eyes and ears alert, no headphones, no music. I scan the horizon, look close and far, stop, zoom in on details for as long as possible, look until the building blurs, and then refocus. The voice in my head narrates and annotates what I am experiencing. Next time, I will go a different way. Even if I follow the same path twice, I will try to look for things I didn't see the first time: spaces between the buildings, the symmetry of the windows, the texture of the paintwork, and the tones of the people passing.

Sometimes, I submerge myself into a crowd, try to identify specifics about the individuals from the mass, and practise navigating a path between all the other bodies without touching – anticipating each organism's movement and predicting the congregation's current.

I spend time thinking about where my breath is placed and where it is sitting in my body. Is it high in the chest or low in the stomach? Is it fast and tight, or slower and free-flowing? And I conscientiously check the alignments in my figure: are my knees tracking over my toes as I walk, are my hips and shoulders evenly balanced? Is my head floating like a helium balloon on a string? Are my eyes easy in their sockets and my jaw relaxed? By quietly posing questions to myself about my body and observing aspects of the sensory feedback, new forms of awareness emerge. In cultivating this particular form of attention, this listening body, both internally and externally, we can create a menu of action that transforms us and reframes our future capacities to sense. As we reconnect and revive our sensorial pathways into awareness, we activate entirely new states of physical understanding.

Take a snapshot of me right now: I am typing at this keyboard on a desk slightly too low for me. I notice that the top of my back is curved, and my head stooped. I am peering into the screen as my chosen font size draws me forward, and I hold tension in my neck. My legs are crossed under the table and resting at an odd angle. I'm hot, and the screen is too bright, causing me to squint and breathe shallowly. And that's just a quick audit. Over time, repetitively in this position, in this constellation of experience, I will entrain these physical signals into an increasingly unhealthy default mode. These maladaptive, learned motor patterns will later become prime instigator of chronic pain and musculoskeletal degeneration. By gently placing my two feet on the ground, engaging my core (my stomach muscles), lengthening my spine, stretching for a moment, breathing into my ribs, drinking some water and repositioning myself so that I am looking slightly up (while balancing my computer on some books), I will not only release my immediate contortion and tomorrow's agony: these minor adjustments will allow me to write for longer, help with my concentration and creativity, and make the embodied, embedded writing experience more pleasurable – a big win for such light-touch attentional investment.

Your senses can also receive a jolt when guided by another person. The sculptor Antony Gormley invites me for a walk in Edinburgh. He guides me through the streets, shifting my attention to this and to that. He directs me repeatedly to look up, beyond the eyes forward, instead of the eyes down. How extraordinary the city appears when the architecture meets the sky. An unexplored threshold. No wonder Antony made *Event Horizon*, which has 31 cast iron figures scattered across a city on bridges, tall buildings, and some on the ground, encouraging us to register the environment in ways we have never before.

On climbing Calton Hill, we rest and look. I still do not see it. Antony points out a body of water in the distance, shimmering and reflective – how could I have missed that? – the Firth of Forth, which links Edinburgh with the North Sea. My gaze was originally too close. Scan further, he says, scan further, and sure enough, it arrives. And with it, a flood of ideas, thoughts, emotions, feelings and, yes, inspiration.

Once you have started to form the habit of attending to attention, listening to your body, noticing sensations, and connecting with your environments, you too can begin to hone the skills you have (re)discovered.

Attending to Proprioception

If attention is our anchor to our physical intelligence, then proprioception serves as our body's guide. While awake, our externally oriented senses – those famed five – allow us to see, hear, touch, smell, and taste. But our 'sixth' sense, the sense called proprioception, allows us to know what position our body parts are in and what they are doing without looking at them.

Proprioception is fundamental for movement, balance and overall physical performance, playing a vital role in everyday activities and athletic endeavours, including dance. You'll be hearing a lot more about it throughout this book.

How many actions could you pull off with your eyes closed? Shut your eyes, put the book down, and pick the book back up. How was that? Can you – without looking – stand up, turn entirely around, and then sit back down in the same position? This might feel risky, but trust that your body's perceptive prowess will ensure you'll not go awry. If that goes OK, can you get up again, fetch a glass of water, and resume your original position? Even if you just shuffled three steps before your eyes sprang open, you'll have clocked how bizarre it feels to suddenly have to attend to where your body is facing and what your limbs are up to from the inside after decades spent relying primarily on vision to position yourself in space.

Now that you've focused on what your proprioception feels like, you'll likely be able to pinpoint other instances when you harness it. You do it all the time – we all do.

Every day, we perform countless acts without looking or attending: eating lunch while glued to a screen, shampooing in the shower, scrabbling around the glovebox for change while driving. While conscious attention is fixed in one place (such as tidying up the living room while chatting on the phone), anything else you do is only possible because your proprioception – literally and figuratively – has your back. This stealth hero sense facilitates every single move we make. But we only usually attend to and through it when we have to do something we're not used to, like walking around with our eyes closed or learning a new dance move.

Crucially, proprioception's under-the-radar mode is a critical adaptive function of embodied attention. Since we can't consciously attend to every sense at once, one is prioritised. In contrast, the others quietly go about their business until they encounter an anomaly (a sensory challenge) or we choose to tune into their feeds.

Although we might flick between vision and proprioception while playing basketball in the backyard or dancing in large groups, primarily, when navigating our bodies through space, we prioritise vision as a means of self-location.

As soon as an action, such as taking a sip of water while holding a book, is deemed functional, it's stored in our personal library of 'unthinking' behaviours as a habitual action or reaction and we quit attending to its physicality. This is because once we're secure in the knowledge that proprioception has water-sipping covered, our minds are free to attend to other tasks, such as reading this paragraph.

Proprioception, then, is an ability that all humans possess and can train. I have a revealing exercise that I deploy with dancers and non-dancers to highlight and test their proprioceptive sense-making, especially how quickly movement patterns can become habituated while relying on vision. Equally, our movement tendencies and biases form helpfully and unhelpfully in all of us. By restricting one sense, in this case, vision, our proprioceptive sense embarks on a new journey of discovery. Try this, then:

Place an empty bottle on the floor.
Step back 5 to 6 metres.
Close your eyes and move towards the bottle.
Stop just before you get to it, keeping your eyes closed.
Pick it up.

Sounds easy, right? But if you try it, you'll probably find it far harder than you think. What this game does is highlight how flawed our internalised conception of space is when deprived of visual cues. Some people set off in entirely the wrong direction, and most – biased as we are to the left- or right-hand side – veer away before nearing the target. And given that there's a relationship between the size of the step and speed (again, part of your unique movement signature – we'll talk about this later), it's common to walk too quickly, only to slow as bearings are lost, so that even if we're on course, we wind up short. The missing discipline is an equal, evenly paced step length along an unwavering trajectory.

A further proprioceptive miss is often revealed once players are about to claim the prize. They overcompensate by making an exaggerated scooping gesture with their arms: a contingency grab. With sight, one simply and economically reaches for the bottle and picks it up. When blind, there is a proprioceptive miscalculation (or at least a less economic recovery action). Here, we are reminded that proprioception is also about reading forces with the body – not just gravity and others that act upon it, but also having a nuanced understanding of and control over our own force so that we can apply acceptable degrees of heaviness and lightness, sharpness and smoothness, tension and relaxation to motion.

When Company Wayne McGregor, the contemporary dance company I founded in 1992, was researching our work *AtaXia* for performance at the School of Experimental Psychology, Cambridge, we engaged in a revelatory proprioceptive experiment on adaptation and after-effects, based on the 1950s Innsbruck Goggle Experiments.

We started from the principle that dancers are expert proprioceptive entities. Over time, after much physical training and once challenging movement patterns and behaviours have become functional, they no longer need their vision to navigate groups or to interact in communities of complex physical action. Even when upside down, flying in the air or rotating while twisting, they seamlessly own their sense of self in space. It is not just that they know where their leg, or their partner's arm, will be during a particular phrase; they will also know how quickly and with what power both are moving because proprioception also tells us about reading forces: heavy and light, and sharp and smooth, and tense and relaxed, and bound and flowing. But given their proprioceptive skill level, how do you test and provoke expert dancers' bodies and minds into new terrain by (re)learning movement? How do they 'feel' what it is like to be momentarily unable to be proprioceptively reliable?

During our experiment, each dancer was given glasses that warped or perturbed their vision. These prism goggles (you can buy them online and try them for yourself – and you should!) allowed them to invert and distort their visual field by flipping the top and bottom or left and right. Their world was upside down or flipped sideways.

Initially, dancers froze, their bodies confused by their brains' new signals. Their constructed reality was being challenged, and their once 'natural' proprioceptive senses had become unreliable. Dancers could not confidently put one foot in front of another without stumbling. Tentatively, slowly, arms outstretched zombie-style, these physical experts started to relearn how to walk. As their confidence grew, so did their ambition. The dancers began experimenting and playing with their freshly calibrated locomotion, balance, and orientation.

The fun increased when the dancers started to partner with one another. With their whole relational world visually subverted, a simple handshake gesture was nigh on impossible, let alone a dance interaction with another proprioceptive novice.

Until … until it wasn't. It was remarkable to witness the dancers' movements adapt to their upside-down reality as their brains retrained themselves to match vision and feeling. The dancers interacted, found their spatial anchors, and elegantly improvised together, in unison, in flow, as if they were seeing and sensing as they had before.

When the glasses were removed, the dancers had to repeat the adaptation process to return to their original coordinates. Wearing their glasses had reconditioned their brain–body reality and their proprioceptive sense. They encountered the same dyspraxia and had to make the same efforts to relearn basic movements. But, extraordinarily, this time, the speed of adjustment was exponentially quicker. Before long, they could all move back and forth between wearing the goggles and taking them off in an accomplished and proficient way, with almost no disturbance at all.

For us, the experiment's richness emerged from the way in which it allowed the dancers to access innovative movement vocabulary due to the difficulty the goggles presented. Each dancer had entered a place where their body had temporarily forgotten how to move, the signals between the brain and body had been disrupted, and their usual physical responses and visual cues were altered. This experiment not only offered us an empathetic understanding, through embodied experience, of the terror and difficulty such brain–body disorders can inspire, but it reminded us of what we take for granted proprioceptively and how, with nourishment, by shifting attention, we can all be improved proprioceptive adventurers.

Our proprioception and its sensors in our muscles, tendons, joints and skin, our senses and their organs – our vision and eyes, hearing and ears – are only part of the vast perceptive equation. Each feed into specific perceptive (and many other) zones within the brain. And much like our biceps, they shrink if our mental 'muscles' aren't flexed. As brain space is precious and brain regions territorial, at a base level, if we dwell within a proprioceptive comfort zone by never trialling a novel activity, that system, including its bodily pathways and brain space, may erode.[1]

So, in the same way that we look after our other sensory systems to ensure their (and so our) continued well-being – for example, by wearing protective headphones when using heavy machinery or taking regular screen breaks to safeguard our eyes – tending to our proprioception, for example by focusing to and through it to sharpen its perceptive pathways and stretch its range, will serve us well in the long run.

What if, rather than thinking about the body as a thing to optimise, we paid it (and ourselves) full dues by acknowledging it as the mind-blowingly incredible instrument it already is? Our very own bespoke, painstakingly crafted instrument, with all the musicality and soulfulness, versatility and sophistication, and emotional and expressive potential that entails?

Every single body holds all these qualities. We need to learn how to play ourselves better. Rather than totting up steps and calories, inches, and pounds — that is, thinking of our physical selves as lists of quantified stats — what if we commit to accessing and enhancing our unique creative possibilities? We need to discover the many ways that our bodies, and thus we, can express our thoughts and emotions physically. Imagine how amazing it would feel to play yourself like a concert piano as opposed to switching between Casio presets. Imagine the freedom such physical fluency might ignite.

Throughout this book, we'll continue honing the skills of embodied attention you've been building so far by learning how to notice, experiment with, and extend a few of your own instrument's lesser-championed qualities.

Our eventual goal is to afford ourselves new physical possibilities by applying some of the following ideas to your day-to-day activities: you'll begin attuning your physical self to your expressive self, laying the groundwork for future virtuosity.

2

We are Unique

Our personal movement habits are informed by what our bodies do naturally: run, walk, sit, stand, fall, climb, crawl, lift, and fold. These fundamental actions are learned early in life and are designed based on our morphology: our size, weight, limb length, gait, height, strength, flexibility, etc.

In turn, our days are full of hundreds of repeated actions and habitual movements: the well-rehearsed motions of eating, sleeping and working that we have fine-tuned into a particular way of doing things. They might have been designed for expediency or to suit our internal body clocks. These are specific physical habits (how you eat spaghetti, the fetal position in which you sleep, your long, rangy stride) that any loved one can mimic with great enthusiasm if asked.

But precisely because they're possible without concentration or attention, they tend not to stand out in memory. It's harder for us to identify them ourselves than for another person. Some will have been kicking around since childhood, some since your twenties, many from whenever you settled into your current home, some since last month; even yesterday, if you've just altered your routine, been travelling, or taken up a new pastime.

One reason we develop these recognisable signatures is that we want our bodies to perform reliably. We want to be able to catch the ball, cook on the stove, or walk downstairs without having to think about it. Suppose every action begged concentration and had to be configured anew. We'd be paralysed, not least because it takes significant effort to master, or even just progress, any complex coordination sequence.

Luckily, over time, our bodies learn to perform these movement habits 'without thinking': they become, as we learnt in the last chapter, proprioceptive. Our minds and bodies coordinate so beautifully that this gigantic habit trove can be called upon 'unthinkingly', executed on autopilot, and relied upon not to go awry. This frees our brains' higher cortexes to do other things.

These physical and cognitive habits are positive aspects of what makes you you. They are key attributes in everyone's unique physical signature.

Recognising YOUNESS

Humans are hard-wired to recognise these unique physical signatures – to recognise YOUNESS. It's part of the fight-or-flight adaptations we have evolved over centuries. By sight, across long distances, we can recognise if someone (or something) is a friend or foe (or food) just from that person's gait and stance (or shape). We also recognise by sound – the rhythm of their footfall, the space between their actions, and the noise produced by their weight hitting the ground. All our senses can participate in this process. Smell is important. So, too, is touch: some recognise even by feel alone.

In a 2021 Studio Wayne McGregor collaboration with Random International and BWM, *No One Is an Island*, we experimented with how much information is needed for a moving shape to appear human. When arranged and animated, robotically guided points of light form the shape of a walking person. Fifteen points of light are all it takes to recognise biological human action; but even the slightest change in the position of these points can cause the shape to revert to an inorganic, abstract arrangement.

This highlights the human ability to recognise and differentiate movement. The more time we spend with another person, the more detail we recognise in their gestures, the force of their touch, or their physical expression. We can translate these cues, this detailed physical handwriting, into emotional meaning.

Behavioural Codes

In daily life, gestural language is crucial to your physical signature – your youness. It reveals personality, mood and a lifetime of movement experiences and social interactions. How we use our hands to communicate today tells of the physical styles we were surrounded by as we developed, the codified dialects we absorbed as we moved through play groups, school yards, friendship circles, sports teams, dance schools, and professional cliques.

How much freedom we are granted and we have to express ourselves also plays an instrumental role in shaping our physical signatures. If, for example, we were raised in a community with strictly defined gender roles, where men neither cry nor use softer 'feminine' gestures, then any male-designated children with looser, graceful physicalities may have elicited raised eyebrows, if not attracted outright bigotry and homophobia. As a result, those children may have tried to adjust their gestural traits.

Whether explicitly or implicitly, we internalise the behavioural codes we dwell among. If we're lucky enough to have been raised in an environment where we've been able to express ourselves authentically, then we might have only needed to mask our physical signatures – to code switch – while travelling through less welcoming places.

Boys Don't Dance – HAHAHAHA

I can speak to this from my own experience. When I was growing up, people would say: 'You're too "gay" [that is, effeminate] in how you move.' Or: 'You're such a poof.' They'd be talking about how I used my head and my hands, the positions my legs took (whether closed or crossed), the expressivity of my arms and the frequency with which I blinked – apparently, even my blinking was too soft or too frequent for some.

I was lucky that I was always supported at home: nobody there would ever tell me to be more manly or less expressive. But it did affect me at school, and I learned to play a role of masculinity that I believed was expected of me. You do the physical equivalent of deepening your voice: you 'harden' your stance, play football and try to be less fluid in your gestures – it's restrictive and inhibiting. In building these other presentations – physical and verbal languages – you freeze parts of yourself until you're in a place where you can 'relax'. But you're bluffing! And you are damaged by that. With all my experience and expertise (or because of it), I'm still mindful of how my body reads – especially initially. Yes, I can code shift when appropriate or necessary, but I am also aware that I code shift because it helps build rapport with the other person. People have biases and prejudices; I can't change that by myself, so I use my skills to adjust my physical signature to the context. We all do this to a certain degree, but the authenticity of self, expressed through and with your body's genuine physical signature, shouldn't be altered beyond recognition.

Humans learn early to curb their physicality as a protective mechanism. To not stand out in case they are perceived as a threat, to brace themselves against false accusations of aggressive behaviour – or worse – as a reaction to implicit bias or flagrant racism. A toxic layer of codification is often embedded into the physical self.

Social ideas of gender can shape our expressive behaviour and physical signature, for example, by shaming us into 'manning up' or acting more 'lady-like'. But it's not just other humans that can impact how we express ourselves physically, socially, domestically and professionally.

Physical gender 'norms' are also absorbed through the actions and games children play with from infancy. Male children are often given things to build or construct, drive or fly, knock about or blow up; they learn about physical forces, such as gravity, and how bodies travel through space. Imagine what movement patterns a rocket ship might prompt as opposed to, say, a doll house.

Indeed, female children are sooner given things to decorate, dress, craft, and care for. They're expected to interact with objects thoughtfully and expressively (apart from when it comes to noise), with attention given to artistic detail. All the while, the skills they absorb are abidingly aesthetic and emotional.

Boys, meanwhile, design and destroy worlds, make rowdy messes, throw themselves around, and bound off surfaces pretending to be superheroes: bravo, son! Girls keep their things (and themselves) calm, neat and pleasing to the eye, and obediently stay within lines (and confines): perfect, darling!

In later years, this parlays into physical thinking biases, as do gendered notions around who is and is not allowed to display agency, effort and self-confidence, and which emotion is meant to be shown by whom. Whom is fear supposed to affect? Who's traditionally raised to be afraid – and who's raised to feel ashamed if they get scared? Yes, these are outdated tropes. And yes, progress is being made as we awaken to the limits we place on our kids. But these stereotypes remain ingrained. Still, it's never too late to unpick them from our (or our children's) physical signatures so that all may express themselves freely.

Physical Signatures – YOUNESS

My baseline physical signature, like yours, is fundamentally conditioned by my cultural, domestic, and prosaic experiences, which, as we have acknowledged, are constantly changing and evolving. Some aspects of my physical signature will now be entrained habits (both efficient and inefficient, positive and negative – more on which soon), and other parts will be susceptible to code shifting.

Specialised input into my physical signature comes from the additional somatic experiences I have acquired and continue to acquire during my professional dance training and creative life. These experiences become traceable attributes of my movement handwriting. Any specialist somatic training, whether rock climbing, paragliding, or deep-sea diving, will leave its trace on your physical signature.

For over eighteen years, I danced professionally, in my own company, in my own choreography. While performing worldwide to audiences from Azerbaijan to Medellín, I developed a distinct, idiosyncratic physical vocabulary based on my particular morphology and various important dance influences, including the rave and techno culture of the 1990s.

With great fluidity, changes of initiation and direction, and an ability, learned through body-popping, to articulate small parts of my body in rapid sequence to generate a stream of eddying movement, I carved out a singular voice in the burgeoning London independent dance scene. You could say I had a unique physical signature and that my individuality was championed in the dance world then.

Although I had received a first-class honours degree in choreography from Leeds University and experienced many techniques throughout that course, I never embarked upon a formal training programme in any one style. More like a magpie, I sampled and discovered many styles and proficiencies, fused them into an incomplete whole, and then smashed them with my own. An evolving hybrid – free from restraint – free from everything!

As I write this, I realise that such physical hybridity is not unusual. It is commonplace. All of us, including the dancers I'm attracted to as performers, have rich physical histories. Such dancers come from various parts of the world and carry with them, in their bodies, a particular perpective. They may well have trained in a known dance style, but they've amalgamated this discipline into their bodies nuancing their personal physical signature.

A Dancer's Physical Signature

Dancers trained in the technique of ballet will have a particular set of physical tells and embodied experiences reflecting through their body: turned-out legs and feet, vertical posture, excellent alignment and elegant coordination. Contemporary dancers, depending on their given technical training, might demonstrate incredible floor skills (from the technique of flying low), use of weight and contraction (as seen in the work of Martha Graham), or improvised partner work (learned from American West Coast swing) – the more of these techniques that reside in their signatures, the more range and flexibility they have as dancers.

Each of these areas of skill will leave an imprint, and each will be recognisable when the dancer dances. That might be in the way they hold themselves, the way they move, the kinetic pathways they enjoy exploring, the value they place on speed, variety, direction, weight, and floorwork, including aesthetic decisions: the key dance phrases and gestures they like to perform, and the ones that they think are virtuosic and worth repeating.

These play out repeatedly in their improvisations; somatic patterns that are so ingrained that even when given a novel movement task, they emerge in various transparencies throughout a dancer's body. For example, a dancer in Company Wayne McGregor named Salvo enjoys and excels at floorwork; he naturally finds his way to the floor on every creative exercise and makes incredible phrases of movement close to the ground. The movement is also fluid and elastic – gummy-like. Another, Hanna, conversely expresses her power in micro-movements, a series of sharp, fast, unexpected exchanges and flicks in the body. She is more vertical and moves less through space. This is her recognisable 'default'.

The more dancers train and gain experience in various technical (and creative) processes, the more these transparencies update themselves. They become layers of experience, a living library constantly adding to itself, a thick stratum upon strata of tacit knowledge held in the body's tissues and fibres. Some call this muscle memory.

Dancers' stylistic training, performance experience, and distinct combination of habits are precisely why they're selected for a role or invited to join a company. Their unique approaches to movement sets them apart, either by contrasting with or complementing others. Yes, this uniqueness is evident even in pedestrian actions like sitting, walking, turning, or stepping. Each person performs these actions in their own recognisable way, but more remarkably, their individuality is as visible in the broader movement vocabularies and dance-related genres they have ingested.

Dancers can be reliably asked to move in a particular way: their way. Individually identifiable physical signatures are critical for dance-making. I always cast dancers with a distinct way of moving and a stylistic handwriting of their own (as long as this is balanced with an open-minded, adventurous spirit!) Their movement habits can be harnessed (and challenged) in a stimulating, creative process. If Salvo is in the rehearsal, I can creatively employ that elasticity and ability to invent something new. If it's Hanna, I can leverage her biases to innovate in another direction – the same but different. If I partner Salvo with Hanna – understanding their entrained physicality (one preferring to be low and one who likes to be high) – I already have an interesting motion dynamic to explore.

Re-shaping Habits

Our physical signatures and movement habits help identify us and allow us to do things effortlessly and quickly. When these efficiencies operate in a well-calibrated body, the effect is impressive. However, trouble arises when we modify these efficient habits (or learn them awkwardly to begin with), forming potentially harmful patterns. You might have learned to walk by putting all your weight on the outside of your foot and thus ended up with a kneecap issue. Or it could even be that something as minor as wearing a certain kind of jean that pulls your knee round has led to the same niggle. Either way, over time if you don't notice and remedy the issue, your alignment will be affected as your muscle starts to pull, twist and turn other areas of your body – a bias in your unique physical signature.

One of our most obvious biases is how we allow one side of our body to dominate, depending on whether we are left- or right-handed. If somebody throws you a ball, you'll catch it with your dominant hand. It determines how we open doors, reach for things, and even get out of a chair: most people don't get up off a chair evenly; they push down on either the right or left side. (This is why when you're conducting rehab for older people, you make them practise both.)

Habits make our lives easier in many ways and limit us in others. But you can do other things to revitalise your body and drag it out of its practised routines.

It helps to take a step back to identify the bias, the habit(s) causing you problems. Nothing exists in isolation. Every physical sensation, good or bad, is unique to your body and its movements (chronic and otherwise), as contextualised by your day-to-day life. Lifestyle, posture, ergonomics (the study of you in your working environment) and sleep position greatly influence physical signatures. Contextualising your physicality allows you to unpick habits if they're not useful – and celebrate them if they are.

★ ★ ★

For instance, seasonal changes affect our physical signatures. When it's freezing outside, people scuttle around, eyes down against the bitter wind, shoulders hunched, stiffening arms crossed, or hands jammed into pockets. Come spring, bodies open out, faces light up, and eyes meet again – warm climes equal warm vibes! Additionally, on a chemical level, the physical discomfort of being cold alters our social behaviour, making us less trusting and more avoidant of others, making them feel colder and becoming less trusting – a physical echo chamber! All of which is to say that our physical behaviour – down to every detail – matters. Ultimately, there is no such thing as a throwaway move.

Imagine you've been hunched over a desk all day, and then go out for a run. What you've been doing for the eight hours before you set off will be embodied in your stance before you've even put on your trainers. You're already going to have forward motion in your shoulders, which will leave you slightly bowed. And that's going to have a specific effect.

That's why a lifestyle awareness audit is useful. It allows you to think about the big (and small) physical movements in your day and their impact on your physicality. You can reshape and recalibrate your physical signature through attending to these habits. As you read this, get up, or lie down and stretch properly. Luxuriate in it! Just be sure to pay full embodied attention to what your body is saying as you do. Try not to think about anything at all beyond the sensations you experience. Feel for anywhere tight, floppy, crunchy, spiky, gooey, sticky, numb, or just hurt.

Could you make a note of what and where those feelings are?

Next, visualise what a standard weekday – or even just yesterday – looks and feels like from an embodied perspective. How much time do you spend stretching, dressing, standing, commuting (think of what contortions your body is forced into when you're crammed into a packed train), eating, sitting, twisting, typing, stool-perching, shopping, driving, cooking, kid-wrangling, running, phone-gazing, lounging, sleeping?

While these activities are taking place, what is your body's position? Where is your weight held – and for how long? Is anything you are wearing restricting your movement?

Scan your body, focusing on the cranky or tense spots you noticed during that stretch. Which of those can you attribute to certain habitual bodily positions, actions, poses or even items of clothing?

Only some habits will need immediate attention. Some will be unimportant movement habits, but others will have the potential to cause harm, or will already be doing so. The issues come when a habit no longer serves our physicality (for example, when we attempt to run at age fifty with the stride we adopted at age fifteen), when we adopt actions in an imprecise or ill-instructed fashion (such as lunging with misaligned ankles, knees and hips); or when we lazily transfer a habitual action to an alternative application (for example, stooping directly down and then reaching straight up, rather than crouching and rotating spine and limbs, to do things like fix under-sink pipes), which can wreak havoc on an inflexible back.

If we fail to notice these maladaptations as they occur, our body, in its eternal drive for efficiency, will absorb habits that may corrupt our entire form.

As you conduct this audit, the other thing to remember is that the body is forged of interdependent systems: muscular, skeletal, circulatory, and so on. These relations are causal; everything affects everything else! There's always a relation between tension in one place and its expression in another; a domino rally of niggles and aches that, once triggered, becomes increasingly difficult to trace back to the source. If shoes hobble toes, our constricted steps reverberate through our arches, ankles, calves, knees, hips, back, and neck. If we only ever carry our laptop on one shoulder, the neck, head, ribs, spine, hips, knees, ankles, and feet will feel the lopsided strain. A runner suffering from discomfort might go into a store for a shoe test and be convinced that their new pair of trainers will be sufficient to fix their instability or alignment issues. They don't apply the information from the shoe test across the whole ergonomics of their body.

A dancer's habituated physical tendencies can cause issues, too. For instance, some dancers have developed poor technical habits that have become ingrained. An example would be a dancer consistently sinking heavily into their hips when performing a pirouette to the left or misaligning their knees when jumping, potentially leading to injury over time. These habits are severe, usually dealt with and worked on in the daily class, as well as physiotherapy and strengthening regimes that underpin all dancers' lives. Regular maintenance of the dancer's body helps identify issues with a dancer's core technique early and address them.

In the past decade, there has been a significant enhancement in professional dancer training, particularly in the application of sports science to support dancers' body maintenance, injury prevention, and rehabilitation. This development has been instrumental in transforming perspectives on dancer health and training by utilising data-driven, research-based approaches to treatment. Physical habits are measured and assessed using the latest technology, and appropriate programmes are implemented to counteract them.

New Sensation

While the body's technical proficiencies, which we acquire from birth and develop as we grow, are essential, what happens when those habits start to inhibit doing? What happens when habit reduces our options and the body and mind grow bored? Our reliance on these routine, reliable actions can blind us to what our bodies are capable of. They can shut us off from learning and experiencing new things.

Think how activated we become when we break away from our routine activities and learn a new sport, discover yoga, or take that first salsa class – no wonder we are re-engaged when our lover diverges from the usual map and explores virgin erogenous territory. Our bodies might be perfectly capable of repetition, but they also seek and need new sensations and new pathways of experience that can lead to enriched physical living.

Novelty in the body is the opposite of habit. It's essential to allow yourself to experience novel things: learn to krump! Give whitewater rafting a go! Plunge into freezing cold water! But also think about the potentially transformational work of changing, refining, or improving the physical and cognitive habits you already possess.

With this in mind, you can think about breaking or disrupting old, damaging habits and replacing them with something new. It could be something as banal as your twisting habits. If you're going to open a jar, do you have a twisting habit? Which hand turns? Can you be on the other side? Do you anchor and twist, or twist the bottom off the top? Even just doing it with the other hand enlivens something in you that wouldn't have lit up otherwise, and it might mean you stop putting unnecessary strain on your back.

All of this is a process of trial and error, just as it would have been when you were a child forming your first movement habits. Experts in functional movement spend a lot of time looking at toddlers and how they build from their infant movements, like bear crawls, into confident walking, running and climbing. Young children learn by trial and error. Adults can falsely believe that their days of exploring the various mechanics of putting one foot after another are over. We claim neither the time nor the appetite to learn new things by getting stuff wrong. However, our desire for continuity is in constant tension with the world's uncertainty. If our gaits are programmed to manage the length of just one set of stairs, what happens when facing an irregular set?

Somatic Adventurers

In dance, you'll find many people committed to facing an irregular set of motions – that is, to pushing the boundaries of their somatic (anything related to the body and physical feeling) experience, looking for sensations they've not experienced before. They might be doing that through experiential frameworks in a body that they're trying to unlock or through cognitive frameworks, which push them into a new direction.

But we can all become somatic adventurers. It's liberating and freeing, both physically and mentally. Living an exploratory movement life is significant at any level of adventure you choose. It means you're not stuck in repetition; you're not bored with yourself or your interactions. Rather, you are stimulated and creative. It's a fundamental truth about human nature: if we can stay in the learning zone through challenges, we lead richer, more embodied lives. Every risk we take and every new activity we attempt are vital experiences that foster courage, tenacity, improvisational skill, and creativity.

It's why I want the dancers in my company to have as varied a physical diet as possible. I'll bring in a capoeira expert to teach them, a hip-hop dancer, or artists from a broad range of physical disciplines to inspire the body to tackle new things. These experiences change, extend and recalibrate what they can do with their own bodies.

All physical inputs change the way you move. You might have a singular technical thrust, such as ballet, Limón, Bharatanatyam, or jazz, but even staying within that discipline's parameters, you can go deeper, experience more, or hijack other dance forms and competencies and explore them in a combination of interesting ways.

The more you dance and experience dancing, the more adaptations and range you add to your physical handwriting. Dancing is a contract with yourself to change! Not change different, but change, and... it's addictive. Developing one's physical intelligence is a process of lifelong learning. How might you enhance your unique physical signature, your uniqueness, through adventure?

3
We are 3D

A work day – I'm heading into central London, and as I leave my front door, I face a road surrounded by fields. No high buildings are nearby, and I can see the cityscape in the distance. The sky is vast; I am small. As I walk down the hill, houses encroach on either side, and the pavement compresses. I feel narrower now, passing people in linear slides. Approaching a busy crossroads, the taller buildings loom ever more prominent as I near the station entrance. Soon, I, too, am in the crush of the crowd. Navigating the human maze, dodging and jostling, I realise I feel narrower and shorter with opposing forces on all sides, like a walking pixel.

    I know this won't last long, and I'm eager for change as we finally spread out along the escalator into the tube. There is a moment of pause as I'm moved downwards with little chance of the unexpected. We wait on the platform, too close to each other as we shuffle along, avoiding the edge and facing a wall. With zero possibility of moving straight ahead, we're forced to think laterally – our future space is now side to side. The train arrives, and we breathe in, injecting ourselves into the living mass as if we've become a jigsaw piece in search of a slot. The space is tight and claustrophobic. Again, we're being moved by external forces, and our bodily agency has diminished: there is no way out!

Soon enough, the door releases, and we exhale as we clamour for platform space up front. We have height again but dare not look up as our eyes dart between bobbing heads in front. Until, finally, we're outside again. Free to move as we desire – free to breathe!

Perception of Spaces

We continually reinterpret the spaces around us to suit our purposes, to better align with how those spaces make us feel, and thus move. We navigate our environment by constructing our own version of it, shaped by the elements we focus on. Our desires and needs prime our attention, as do our perceptions, movements, gazes, hearing, sense of smell, and more. If we are hungry, we'll search for a snack bar in the tube station; if exhausted, we'll look for a quiet spot at the far end of the platform before the tube arrives. If we feel threatened, we'll be more attuned to viable escape routes and stand close to the stairwell. These spatial 'affordances' mediate our mental representations of the places we traverse. Spaces, although arguably neutral, are not neutrally rendered in our minds; at the base level they are either attractive or repellent.

Movement potentials within spaces also influence our perception of them as our personal opportunities for action within them affect how we comprehend our location. The long flight of stairs at Angel tube station seems even steeper to an exhausted, thirsty commuter geared up to conserve energy. A child reads the gigantic broken-down escalator with a joyful 'weeeeeeeeeee!!!' Mum reads it as a trip to A&E. And we're all aware of how much mood tints the environment. On an extreme level, for the clinically depressed, even on a radiant day, the world is bled of colour. This way, we reformulate the space around our bodies into meaningful spatial conceptions, highlighting movement possibilities.

By taking you on that commute, I shared how I tend to do this and how it feels in my body. Of course, you would have a different experience.

Our ability to mentally refashion spaces as we move through them is then determined by what we choose or are able to perceive. This includes how we choose or are able to perceive our bodies. Once again, our physical intelligence comes into play here.

Body in Space

Think for a moment about your own body; try to bring a visual image of it into your mind. Perhaps you called up an image of yourself in a photograph or a memory of yourself standing in front of a mirror. If so, that's not surprising. We're so used to seeing ourselves depicted or reflected upon two-dimensional surfaces that it's easy to forget we have depth, volume, sides and a behind – that the body exists in three dimensions.

WE ARE 3D

We all have an intellectual sense of our body as something that exists in three dimensions. But in practice, our experience of our bodies is mainly flat and two-dimensional. We don't have an adequate mental model of the whole of our body. We don't know what we look like from behind, partly because so much of our sense of our bodies is mediated and re-represented through flat objects, like mirrors and screens. We often feel like flat things as we walk through the world – a paper-cut-out version of Leonardo da Vinci's Vitruvian Man.

Much like hearing your own voice in a recording, when we experience our bodies all the way around, we are unrecognisable to ourselves, which can come as quite a shock.

Imagine one of those reality makeover shows where the participants are confronted by a wrap-around mirror for the first time in their lives, providing a complete view that allows them to see their proper three-dimensional form. Their reactions are always the same – awe followed by horror! A similar schism ensues when you volumetrically capture a star actor for CGI scenes in a movie and build their avatar in 3D. Even though their human measurements are rendered precisely, with infinitesimal accuracy, their pain at seeing themselves 'in the round' is palpable. Again, this isn't surprising; we rarely think about our physical picture from behind, our particular curve in the neck, our placement of the head, the space between our legs, our width, our gait, our 'fit' in the world from this perspective. Our faces and fronts most prime us, and thus come to form our de facto image of ourselves.

Quick exercise: take a few photographs of yourself (or ask a friend to) from behind – in an everyday conversation or activity. Study it. What do you notice about your physicality from this perspective? Note it down.

Dancers are no different. You might imagine that dancers would always think and dance in three dimensions. Very often, they don't. This is undoubtedly the result of their endless time in front of mirrors. They're trained, rehearsed, and created upon watching a flat image of themselves in mirrors, which is a dominant feature in the front of any ballet studio. This flatness is reinforced when they see photographs or films of themselves on stage – a series of two-dimensional images standing in for the three-dimensional embodied experience. It accustoms them to the concept of their bodies being less multi-dimensional than they are and conditions their mind, subliminally, to a version of self that negates their 360-degree power. (This is why we built Studio Wayne McGregor *without* permanent mirrors in the rehearsal spaces – to free up our dancers' conception of themselves in space.)

The result of years in front of mirrors in the studio is that some dancers can and do dance flatly. A dry presentational and frontal version of themselves moving between two glass planes. No twist, bend, nuance, incremental orientation, angle, or focus differentiation – a fully forward facsimile – a live 2D performer in a 3D world.

I have seen this version of flat dancing many times during my career, visiting even the best dance companies and dancers in the world. Elite performers, phenomenal physical and expressive athletes, experts in motion quietly forgetting that their bodies occupy and enliven spaces three-dimensionally. And it is always a great pleasure to see these dancers release themselves from this stricture by, in the first instance, (re)imagining themselves, (re)minding themselves that they have volume, they have mass, that their three-dimensional self has as much potential in the backspace as they have to the front. They are both bodies in space and have space in their bodies. Through active visualisations and exercises, we encourage a gentle reframing of the historical narrative, namely that your body exists in three dimensions. Suddenly, it's as if the glass planes are removed, and finally, the dancer finds their dance – they occupy a volumetric space with their volumetric bodies, and their bodies speak a new language. Even thinking three-dimensionally makes all the difference.

Let's return briefly to the visualisation from Chapter 1. This time, you try to imagine your body rendered in three dimensions. Picture yourself in the position you are in as you read this book. Mentally hover outside of yourself and travel around your form, considering how your limbs, feet, hands, torso, head and neck would look from multiple angles as they take up and extend into the space around you. Try to notice details, much as a sculptor tasked with casting you in marble might.

Next, imagine a room with an ample empty space that is warm and inviting. It slowly fills with beautiful smoke – a colourful fog. As you walk, run and, yes, dance through the room, the smoke displaces, leaving an imprint of your form behind. You can see it – the negative space where you recently passed. You can walk around it. Walk around the movements of yourself.

Now, move your attention to the inside of your body. Imagine your rib cage, arms, head, and pelvis as containers of space. They, too, fill up with the smoke – micro-spaces form within your body, your body creating negative space within a room. And so it goes on. Quite remarkable, isn't it? How every living moment plays out from within a three-dimensional body as suspended in 360-degree space, whether alone or surrounded by other beings and objects or atmospheric forces?

Feel the change? It is not just dancers who can benefit from this re-imagining of how they exist physically. Once you are aware of yourself as a three-dimensional entity, it changes how you feel about yourself in space.

If we only pay attention to what's straight ahead, straight up and down, and straight to the right and left of us, we end up living in something resembling a floating crucifix. We do not notice or attend to the remaining space around the body: to all intents and purposes, it may as well not exist.

As a result, we won't think or choose to move through it until circumstance forces us: the moment we snatch up falling glassware or catch ourselves as we slip backwards while jogging downhill. Because whether we attend to it or not, that space is always out there, around, behind and beyond our bodies.

When we force our body (or someone else's) to unexpectedly move along angles we habitually ignore – for instance, diagonally and down – we make it vulnerable. I once endured an unnecessary back injury caused by someone anxious to help me with my suitcases who unfortunately had little conception of the body as something existing in three dimensions. I was carrying two heavy trunks up a flight of stairs, and he offered to help. I was managing fine, and declined. Regardless, he decided to yank the suitcase that was in my hand furthest from him, across my whole body towards his, and in the process, twisted me so weirdly, violently, that I had to spend the next two weeks on my back. If only he had been thinking three-dimensionally, he might have audited my body differently, perhaps by stepping back and exchanging places to approach from the other side and gently taking the case into his palm. Or, instead of reaching forward, across and down, he might have effortlessly taken the case closest to him. Next time!

If our movement library is filled with rote (straight ahead, straight up and down, straight right and left) actions, when we have to swiftly swoop, crouch and lift outside of that crucifix-like comfort zone, perhaps to pull a curious toddler back from a pool edge — we find we do not possess the requisite 360 agility. We are out of practice. This is why we must teach our bodies a healthy array of three-dimensional options for when we need to suddenly catch, snatch, reach, twist, tumble, yank and swoop outside of habitual patterns.

Action Patterns + Movement Schemas

It's important to highlight a key aspect of our physical intelligence that helps us understand how the body learns.

For every new action or action sequence, we figure out until we attain proprioceptive reliability – in other words, to the point of being able to do it 'without looking' – our bodies capture a physical imprint of that procedure. Let's call this an action pattern. Simultaneously, inside our skulls, the cognitive plot of that same action is filed away. This mental counterpart of our physical actions is known as a movement schema.

Together, these action patterns and movement schemas form the very foundations of all movement learning. By being encoded into body and brain in a kind of physical archive, we're able to progress to the next stage of our chosen activity, whether that's break-dancing, showjumping, bread-baking, knitting, or navigating the crowds in the tube station.

Once we grasp any action, we can perform it on autopilot while attending to something else entirely – it can be classed as fully proprioceptive and stored away as a movement habit.

Lunchtime – and you set off on a reviving stroll through the park. After veering off into a wooded area, you face a roughly human-sized gap between trees. Still, somehow, you know you won't be able to squeeze through without your coat snagging on the branches. Not long after, you come across a gate and spontaneously leap over it, feeling buoyed by the sunshine. How did you know you couldn't get through the gap and could manage the leap, seemingly without deliberating it? It's thanks to your body schema.

★ ★ ★

When we absorb actions into our movement repertoire by creating movement schemas (that mental map of an embodied action pattern), our brain also contains a schema for our entire body: a body schema. (Note that this is not your body image; that's a whole other thing.) The body schema is critical to your functionality as a walking, gate-leaping human being. It's also the bedrock of your self-conception. As far as physical intelligence attributes go, it's in a class of its own. Yet, much like proprioception, it gets nigh on zero airtime.

What is a Body Schema?

For starters, much like a personalised air traffic control centre, your body schema plays a massive role in green-lighting (or blocking) your every move. An excellent way to think of body schema is as your very own movement control hub, which, while located in the brain, is concerned with where your body parts are and what they're up to every millisecond of every day. So, as you may have guessed, proprioception – how your body parts, individually and in any combination – is hot-wired into body schema, so that it can assist you in making a judgement call on what actions you're actually capable of. For example, that fence leap would have received a hard no if you'd run a marathon the day before that sunny park walk and had duly stiff and aching legs and feet!

Critically, your body's action portfolio, encapsulating every move ever made, includes a take on how you'd fare doing them now. Even if you haven't ridden a bike for twenty years, your body schema would give you the green light to give it a go, as a 'bike-riding' bodily action pattern and mental movement schema are still stored inside.

But you'd also be keenly aware that — as it's been twenty years — those faded imprints need updating to match your current physicality. Thus, you'd naturally be acutely attentive to what your body is doing and how it feels once back in the saddle and instinctively steer clear of the BMX tracks.

We can then define our utterly remarkable body schema as a live, automated cognitive map of our physical condition and action potentials in their totality — although it's perhaps more fun to think of it as our personal air traffic control hub. While the body schema mostly functions under conscious attention, its signals become accessible once we learn how to listen for them.

It makes sense, then, that dancers or indeed experts in any physical discipline, who have evolved sophisticated body schema (an extensive physical+cognitive archive) through decades of training in a range of movement techniques, can access a broader range of action potentials than those who have explored their body's intelligence less. The more varied input you imprint and store in your body's action portfolio, the more potential variation your body can perform and recall, even if it is out of practice.[1] The trick is to open yourself up to the input in the first place.

For dancers, this pursuit of nurturing one's body schema is inherent in their lifelong learning as dance artists. Accessing and accomplishing new forms of dance and breaking physical habits through experimenting with the body's ability to learn unexplored physical languages should be hard-wired into the daily life of a dancer.

However, as we have seen earlier with flat vs three-dimensional dancing, even experienced movement experts need to update and refresh their repertoire. By shining a light on this attribute of physical thinking that usually operates in the background, under the radar, tuning into our body schema, and bringing it to our conscious attention, we can extend and augment it. We want to develop new movement habits that can be stored in our air traffic control hub. We want to transition from limited action portfolios to expansive ones and blast away our habituated examples to be replaced by a plethora of agile options. We want to provide a more decadent space where our bodies can think and behave differently.

If we are to release ourselves from the poverty of the two-dimensional version of ourselves moving in the world by rote, we might need to practise the alternatives. Let's invest in updating our body schema in relation to three-dimensional experience and break our habitual cognitive patterning of flatness.

Spaces We Carry

In the 1920s, the modern dance choreographer Rudolf Laban, who worked tirelessly to democratise dance by making the art form's expertise available to as many people as possible, coined the term 'kinesphere' to describe the imagined sphere of motion that immediately surrounds the human body. As Laban demonstrated, using detailed illustrations, wherever a body goes, their kinesphere travels, too. Outside of our kinesphere, there is general space. With one foot planted, our kinesphere's periphery will always be found at the limits of our extended limbs. Its weight shifts as ours does. All points in the kinesphere can be reached by uncomplicated movements such as bending, stretching, twisting, or by a combination of these, and utilised to expand the zone of reach for a human body.

As a choreographic construct, the kinesphere was sometimes depicted as an icosahedron (a twenty-faced polyhedron), but you don't have to imagine it like this for yourself. It pays to (cognitively) build a personalised visualisation of your kinesphere. For example, encase your body in an imaginary Zorbing ball, or a planetarium ablaze with constellations you can reach out and touch, or a 3D version of the Vitruvian Man. The exact shape of your kinesphere isn't essential, but what is significant is the concept behind it: wherever we find ourselves geographically, our movement potentials are eternally 360° up and down, in front and behind, side to side, diagonal to diagonal.

We often use variations of this principle in the studio with both professional and non-professional dancers as a dynamic testing ground for practising and expanding the variety of our spatial pathways. This increases the opportunities for agile, balanced, coordinated motion and bolsters our habitual movement library so that we develop a new sophistication in the body and visualise it as a three-dimensional entity. We also use it for novel movement generation and inventing unusual dance language, far outside the conventional dance phrases one might expect.

One variation we use for young movement students invites them to stand upright and imagine a cube around their whole body. The limits of the cube are at the limits of their reach. The top of their cube has four corners, and the bottom, too. In the middle of their body, we ask them to imagine a flat surface dissecting their torso, a plate floating mid-way between the top and the bottom of their cube with the same proportions. The body then has three planes – a ceiling, floor, and in-between plane. Want to try?

Using your left hand, and without moving your feet, reach in front of you to the top left corner of your cube. Swap hands and reach the same corner now with your right hand. Touch the opposite front top corner. With your right foot, reach for the front left corner at the bottom of your cube and then the right-hand side with the same foot. Easy, right? With your right elbow, now touch the left front corner of your middle plane, swap elbows and touch the other front side. You are reaching forward through each of the planes.

Now, try all that again, using only the back corners of your cube, the ones behind you. Hand, leg, elbow – it's not quite as effortless and more demanding of your core movement skills.

Next, try numbering each of the corners, going clockwise from the top left – the top plane's corners are 1, 2, 3, and 4, the middle plane's are 5, 6, 7, and 8 and the bottom plane's 9, 10, 11 and 12. Got it? Move your arm into 2, then your leg into 7, then your elbow into 11, your head into 3, your shoulder into 3, your ribs into 5, your knee into 1 – you don't need to keep in the position you started in, you can move between the numbers. OK – more challenging?

Trace a line between all of the cube's even numbers with your finger, then between all of the cube's odd numbers with your forearm, and then try tracing one between 5, 11 and 7 with both your chin and your shin simultaneously. Be creative!

Drawing with your body, started as a primary action task, becomes increasingly complex, with the rules bending and evolving just a little with each iteration. Here, the hierarchy of backspace and front space has been evened out, and the young dancers are encouraged to explore unfamiliar movement directions and patterns entirely in 3D ad infinitum – and remember each action pattern that is performed will have a complete body schema index to go with it, stored forever in mind and body. The more you experience the alien coordinations, the richer your movement library becomes and the more physically literate you will be. It's a physical thinking version of a brain gym.

Maturing your three-dimensional acuity through this awareness and rehearsal of your kinesphere has further real-world benefits.

All physical thinking is interactive; it embraces whatever object or tool we're using at any given time, including pens, hammers, prosthetic limbs, tennis rackets, wheelchairs, and bicycles. Any object we interact with proprioceptively is perceived as a temporary extension of the body. And so, our body schema, centre of gravity and balance, co-ordination, and action possibilities are adapted to accommodate these objects. Kinespheres function this way, too. Remember, their purpose is to aid us in nimbly navigating space, and one of the benefits of thinking about our kinespheres is increased situational awareness.

Essentially, situational awareness grows through the development of our perception skills and the comprehension and prediction of relevant situational cues: for instance, spotting congested spots, interpreting how other people are moving and might soon move, catching street signs or studio lighting cues, and noticing irregular step heights.

This requires that we continually, attentively, scan our 360° environment for such cues, keeping focused on how we're moving. By doing this, we'll stay equipped to spot hazards of any nature in good time to react. For example, if we're scanning in every direction, we'll be more likely to catch – and thus avoid – an imminent collision.

Fundamentally, situational awareness demands that we dynamically control our combined attentional and motor resources, drawing upon our freshly sharpened perceptive and physical skills day in and day out.

It can, for instance, be deployed as a road safety tool. Take a moment to replay the last time you rode a bike. Considering your stance in the saddle, how you usually scan for traffic, and how you navigate it, remember that we mentally refashion spaces as we move through them: what sort of space do you reckon you've been projecting around yourself while cycling?

Do you focus on what's straight ahead alone? Or gaze about 360 degrees?

Do you imagine yourself seated on your bike as narrow and arrow-like, lofty and rotund, or compact and tightly packed?

Do you sit high atop the saddle? Or low, your torso edging over the handlebars?

This may be tricky. If so, think specifically about those moments when you have to negotiate gaps in traffic. If you're confident as you nip between cars, lorries, and other cyclists, your cycling kinesphere is likely slim, long, direct to front, taking minimal account of what's behind, and reliant on speed to gun it through.

All of this is worth considering, because if you cycle within a kinesphere diminished by fear you increase the risks you face. It's safer to expand your kinesphere and thus, mindfully, take up additional room on the road. This generates extra space around your global periphery, ensuring that other drivers or cyclists recognise your presence and providing you with additional time to anticipate their trajectories and react to unexpected situations.

To move confidently across the physical world, the mind needs to be confident that it, in collaboration with the body, 'comprehends' its environment so that it can tell us what is and is not possible. Our body schema only highlights actions we can execute, given our physical capabilities and knowledge bases. The mind shapes and reconstructs the spaces around the body, depending on what we attend to, as guided by our movement habits and present needs and goals. In this way, much as it can project a kinesphere for our body to play within, our mind dictates what is and is not available to us in terms of potential actions. Together, both factors help us decide our action boundaries and, with fresh attention, guide us safely through the next 3D commute home.

4
We are Balance

Walking is all about falling — at least from a conceptual point of view. With every step, there is a lift, and then the foot falls through space, gaining momentum before it hits the floor and repeats on the other side — fall after fall.

There's value in expanding our definition of what a fall is. And to ask: why do we aspire to be vertical? What if instead of thinking of balance as a fixed state, as something to be accomplished, we began to view it as a process? What might arise if we considered balancing a continuous negotiation, which plays out within our mind and body, other humans and objects, and surfaces and forces? If we perceive balance as an endless duet between equilibrium and disequilibrium, what opportunities does that invite?

Our ability to stay upright is the result of ceaseless negotiation. Our muscles, bones, joints, proprioception, vision, and vestibular organs (the balance bits in our ears) communicate with our central nervous system (brain and spine) to help us perceive and adjust our position to accommodate environmental forces and other external factors.

Non-Static Entities

The human body is never entirely still. We are not static entities. Each of us is an organism of perpetual flux and change. However hard it is to strive to fix the body or mind in a solid state, internal motion and transition are inevitable. Like everyone else, dancers can make the mistake of imagining that everything in their bodies has stopped when they balance. In reality, their ribs will be floating, and their feet will be shifting. They'll be making thousands of micro-adjustments to achieve that beautiful pose.

Consider the basics: we breathe approximately 19,000 times a day, at a rate of thirteen breaths per minute. Our heart beats 100,000 times every twenty-four hours, pumping about 5,000 ml of blood around our body. Our skin is never dormant. Our organs, immunity, and metabolic systems do not halt. Even while asleep, our body pursues energetic equilibrium. We know this. But perhaps we've stopped listening to these dynamic flow systems — our inner tides, eddies, and whirlpools.

To access our full physical potential — to tune into our innate musicality — is to accept that there's always movement within, and with that movement comes energy. That central axis of stillness and motion is, in a deeply essential way, our energetic core. Viewed in this way, it's the place from where we ultimately draw the potential to move in any way and any direction. And by connecting with that essential energy at its very source, we can exploit it in lots of dimensions — and with agility.

For example, we can catch the end of a breath to propel a leap, use our heartbeat as a metronome, or surf the sensation of our circulation to dance from deep within. Whenever we need to centre ourselves anew, to reconnect energetically, it is enough to pause, breathe, and listen for our instrument's still point, our body's eternal axis.

This all means that it's genuinely impossible to achieve perfect stasis in any pose. There will always be movement, no matter how slight. And that's a good thing! Once we accept this, we're on track to a far more balanced relationship with balance (in any style, form, shape or guise).

The question is, how do we maintain enough of it to do what we need and/or want to do for as many years as possible? Or, more simply, how can we avoid falling when we'd rather not?

It is critical to understand that balance is a dynamic concept that is always in action. Balance plays a powerful role in the physical thinking canon as it manifests on this active, uneven planet. Balance isn't about maintaining any ideal stance or pose; it's simply what we rely on – our active postural control and stability – while we manoeuvre down a tourist-rammed escalator to catch our train or pick our way along a trail strewn with uneven rocks.

And we can work to improve it. Here again, proprioception – our perception of and control over what the body is up to and where its parts are – is crucial.

Poor proprioception can often be a more reliable indicator of fall risk than issues like failing eyesight. Although both factors can lead to falls, deficits in proprioception more directly impair balance and coordination, which are essential for our stability and fall prevention.

WE ARE BALANCE

Without a decent proprioceptive facility, our body's struggle for basic stabilisation draws resources from our overall bid for surface stability. Understanding the importance of proprioception in improving balance can empower us to take control of our physical well-being.

Our ability to retain a solid range of proprioceptive skills — let alone evolve them — demands stretching their, and so our, limits. Improving our balance means testing and extending our proprioception by negotiating unstable surfaces through our bodies. Ideally, this would involve falling repeatedly. This might seem drastic, but without pushing the body (and the mind) into uncomfortable states, we'll never find a comfortable means of navigating challenging circumstances.

Maintaining equilibrium involves refining our skill at starting and stopping action. Balance is about control — falling is about letting go. We won't know what to control in any given scenario until we understand what to let go. And the first thing to let go of (incrementally, patiently) is the desperate battle to remain vertical. The reflexive flailing that arises from our terror of tumbling only further destabilises the body, increasing the risk of injury.

If, as you're falling, your main thought is, 'Fuck, I'm about to fall', then your body will tense up, which means your landing will be even less comfortable. It's a movement paradox: you must practise the opposite of what your instinct tells you to do. When you fall, the thing to do is give in. Instead of going rigid in your whole body and falling at an entire angle, you soften your knees so that you fall through your knees and your body. You want to be doing something similar with your hands. Rather than pushing away at the surface you're landing on as if trying to fight it off, your hands should receive the force like a shock absorber. It's more a question of crumbling and then falling. There's nothing nicer than having a chewy body – an elastically pliable and robust instrument. It's a lovely feeling.

But to reach the point where you can embrace the thought of falling rather than recoil from it, it helps to think more about why so many people fear it.

Our anxieties around falling are partly a learned response that we pick up as we move into adulthood. As children, we fall easily and laugh, even if hurt is involved. We bounce back up again. Because we enjoy falling once, we do it again and again. Our innate physical intelligence means our bodies know not to tense as they descend to the ground. We fall – going with gravity – we rise, we carry on.

And yet, there's a moment in our development when falling becomes ridiculous and frowned upon. Instead of a fond smile, you get someone shouting, 'Stop messing around!'

The idea of being off-centre gradually becomes a problem; lots of adults are afraid of falling. They hate the prospect of pain, the lack of control, and the effort involved with going down and coming back up again. This is partly because we're taught to minimise effort. In some circumstances, this economy is positive. But it also leads to people doing less. We fear falling, so we restrict our movement and stop exploring how we fall, which means our fear of falling grows.

Anxious thoughts about your balance can mess with the attention you need to perform it adequately. If the fear of falling preoccupies all your attentional resources, the capacity for your senses to support a gentle fall is otherwise engaged. To counter this, you can practice being off-balance to become comfortable with it. Falling will always be easier if you can get rid of tension.

You can begin this process by acknowledging two facts. One is that falls are inevitable. Many of us live in flat-plan cityscapes and use the lift instead of racing downstairs and driving everywhere, but no matter how cosseted we imagine ourselves to be, life inevitably throws a freak upset our way. Pavements are full of potholes. Forest tracks are littered with stray branches. At some point, no matter how careful you are, you're going to trip. The second is that not all falls are equal. Think about what happens when your body falls. For some people, it's going weak at the knees; for others, it's going off balance and falling sideways. Do you fall forward or backward? Do you fall down? Do you fall quickly or slowly? When you fall, what's your reflex? Is your reflex expressed in your elbow or your arm?

Acknowledging these – our biases and tendencies and the relative position of our centre of gravity – helps us construct strategies that help us avoid falling (we realise that we can shift and that the status quo is not the only option). And it is another way to understand more about how our bodies work.

Your Body's Equilibrium Point

Your centre of gravity is the location where the weight of your upper and lower body equals out, serving as your body's equilibrium point. This spot is usually positioned slightly below the belly button for women and slightly above it for men, somewhere between the lower back and the navel. This point tends to shift in response to your posture and body weight alterations. Try to visualise this spot now.

Balance is effectively your ability to control your centre of gravity in fluctuating conditions. The equilibrium point can shift if a change impacts your balance and stability. For example, when a woman becomes pregnant, and her stomach extends outward, her centre of gravity will move forward, too, requiring her to balance differently and possibly feel more unstable. Similarly, a man who does a lot of upper-body weight training and less lower-body training will have a higher balance point.

Where is your centre of gravity? And how aware are you of your centre of gravity and how it changes when you are swinging your arms around aggressively, standing on a ladder, or getting on a bus?

Whenever I'm in a theatre and end up in the 'nosebleed' seats, my anxiety increases due to the sole 'safety' bar that's supposed to prevent a fall from the balcony. The bar's height, resting below my waist, particularly unsettles me. Standing 6'2", the bulk of my body's mass, primarily in my head and pelvis, towers over the bar. Consequently, this setup significantly elevates my risk of accidentally flipping over the edge since my body's centre of gravity in relation to the bar is at my knees. Yet, my equilibrium point is substantially higher. Danger.

In this situation, I rely on a simple physical thinking tactic. Suppose I find myself in a moment of instability, on the verge of hurtling over the barrier (maybe after being bumped by someone else in the theatre). My instinct would be to quickly lower my pelvis to bring my centre of gravity beneath the safety bar, preventing me from toppling over.

Once you've given further thought to how and why you fall, you can proceed to hone your proprioception and postural control. Ideally, you would work alongside another dynamic force. It is always preferable to have something that is inherently unstable, such as another body or an unpredictable surface, to respond to. When you practise alone, you can build many elements relating to core strength, but gaining experience and responsivity is crucial. However, you can create those conditions.

In an early experiment, we collaborated with a group of young free-runners tasked with maintaining balance in an extremely challenging 'instability' environment. Mobile floors moved, vibrated, shook, and slid beneath their feet. From the initial balance chaos that ensued, it was fascinating and encouraging to see how quickly the body adapted to the challenges and how swiftly their bodies recalibrated.

Not many of us can conjure up a mobile floor for experience, but there are alternatives to practise. You might already engage in some of them: surfing, skateboarding, ice skating, horse riding, rock climbing, visiting a pebble beach, free-running, or even kids' ball pools! You could opt for a more slow-paced cross-terrain hike if these sound too 'active'. Alternatively, if you live in a city, you should mimic your toddler as they clamber on and off street furniture. The surface itself isn't crucial – challenge your usual movement patterns so that your body adjusts to unusual positions and forces, stretching your proprioception and enhancing your balance skills.

In conjunction with this, there are more controlled ways of improving your implicit balance. For example, you can work on your essential stability by building strength and mobility in your ankle joints and feet, strengthening your core and posterior chain, or increasing your flexibility so your body doesn't have to accommodate tighter zones.

These can be found in multiple movement practices, such as countless dance styles, yoga, exercise and martial arts. Indeed, if you take up aikido, you'll soon be taught 'ukemi', the art of falling. Realistically, any action series that involves hitting the ground and then getting up, even burpees, will benefit implicit balance.

Another essential element of balance, though often (strangely) overlooked, is our feet. One of the reasons that some of us have problems staying upright (compounded with issues of reduced strength and agility as we age) is that we need a healthy, neutral stance for good balance. Our toes have to spread, the ball of the foot has to be low, the heel relaxed, and you have to have 'floating' knees (imagine that there is 'breath' behind your knees, carrying them gently forwards, and that you are not pushing them backwards into a locked position). You stand in a good parallel position, your feet the same distance apart as your hips, with your toes pointing forward.

Many individuals, including older adults, often have balance problems because this critical neutral position isn't true for them. The crucial point is learning to spread your toes widely and distribute your weight over a broader foot surface. Many of us wear shoes that are too small, tight and narrow for our foot's natural breadth. Shoes that grip your toes too much, narrowing them, will make the foot unstable as it rocks along the road – rendering you unable to use your whole foot efficiently. Don't make this mistake: wear comfortable, appropriately fitting footwear or bare your feet.

You can do many exercises with your feet, which are just about testing where your weight is off-centre. And you can get better at them, steadily reaching the point where you no longer wave your arms around, looking for something to hold on to.

Try to stand in the neutral parallel stance I highlighted above, arms by your side and head tall. Standing straight, slowly move your weight forward (by moving your centre of gravity) over your toes. How far can you go without gripping your toes or stepping? If you are gripping, back up, back to neutral and try again (or slightly lift your toes and replace them on the floor more widely) – this time visualising your toes as a widening surface, a pool of rainwater expanding with the downpour. Try to keep your whole foot (toes, ball and heel) connected to the floor. Feel each part of your foot gently pressing down on the floor. Try to shift your weight forward again. Practise until you can move your weight as far forward as you can without gripping or stepping.

Once you have accomplished this, try moving your weight backward. Take the same notes: the feet should stay connected and relaxed on the floor. How far can you balance backwards without taking your toes off the floor, falling or flailing?

Next, move forward and backward. Your feet are always connected to the floor, but you can now sense the micro-adjustments they have to make relative to your balance position. Once you are comfortable with this exercise, why don't you try side to side (same rules) and then go in any direction? To clarify, this exercise is where you practice moving both forward and backward, side to side, to improve your core balance and augment your off-centredness. When trying this exercise, please keep your 'rectifying' movements minimal – no quick gestures or large waving, as this will only worsen balancing matters. By making micro-adjustments to your feet and overall body position, you will stay upright and be able to steer more effectively. Test how far you can move your centre of gravity without falling over. It's quite a distance!

★ ★ ★

Recently, I attended a rehearsal at Milan's historic La Scala. On stage, the choreographer's assistant worked intensively with a group of dancers. Watching from the stalls, I noticed she was inching backwards, dangerously close to the edge of the stage. Then, in a flash, she fell. Her body dropped several metres into the vast black chasm of the orchestra pit. My heart stopped. A horrendous thud followed. I felt sick in the pit of my stomach. Silence. Swiftly, she was surrounded by the theatre's in-house medical team, and an ambulance arrived quickly. With head and neck enshrined in a brace, she was wheeled out on a stretcher. I was not to see her for the rest of my stay.

But less than a week later, the same woman hugged me in a Berlin theatre bar! My shock revealed my astonishment. She had arrived at the hospital, undergone a wide range of scans and X-rays, and was proven to be totally unscathed. Incredible!

But – how? Apparently, in that moment of falling, that split second of knowing before hitting the ground after a period of accidental free diving, she didn't panic. She didn't think about the fall or fear of crashing into the pit. Instead, she focused on her bodily state, her centre of gravity. In a flash, accepting that she was about to land, she calmly checked her alignment, took a deep breath, and softened her body position and frame, ready to give into the floor. No bracing, no tension. Instead of reflexively resisting, she let go. She made friends with the floor and gave herself the best possible chance to fall well. Remarkable.

This ability to respond precisely, to react sagely, even at bone-shattering velocity, when the natural reflex would be to tense up, is a cornerstone of agility. As is the ability to deploy such reactions with the speed of a reflex.

Everybody is programmed with reflexes, and these involuntary physical responses are a facet of our proprioception. The calibration of this neuromuscular system is what the doctor tests by dinging our knees with a tiny rubber hammer. We've all experienced precautionary reflexes such as yanking our fingers from a searing stovetop – seconds before the pain kicks in. That and other lightning-quick actions are possible because as we develop, our proprioceptors form reflexive loops with our motor neurons, which allow us to act in a flash as the fibres that convey these signals transmit information at 250mph(!).

Our reactions, however, are slower, context-specific responses that we (usually) deploy at a much more considered pace. Yet it's plausible, through attentive practice, to boost reaction times to reflexive speed – to craft neuromuscular loops to support any action series. Indeed, a major component of dance training is the rehearsal of phrases until they're available to the body at such extreme speeds. In this way, reactions become reflexive.

Conversely, reflexes, including those stemming from fears, such as being dropped while being thrown and caught by fellow dancers, may be transformed into controllable reactions via diligent, physically intelligent labour. Once that conversion is complete – even when descending unexpectedly to Earth – time slows down, and options open up.

This was proved not only by the jaw-dropping physical feats of dancers, gymnasts, free runners and aerialists, but also by the rude health, unreal attentional control and exceptional agility of the choreographer's assistant in Milan.

Like riding a bike, falling the right way requires skill and nerve. The only difference is that whereas most people will never forget how to cycle, they still need to remember how to fall.

5
We are Co-ordination

I have often found myself in awe watching someone touch-type. On the one hand, it's a goal- and efficiency-driven practice. On the other, it's dazzlingly virtuosic. Playing the piano is similar, but different. It involves extra dimensions: the interplay and counterpoint of the two playing hands, the use of the pedal, the dynamic range, the expressive and emotive qualities that piano performance requires. Both activities involve phenomenal levels of accuracy, timing, and spatial understanding. They require sophisticated levels of complex coordination, as if the hands were dancing.

Complex physical tasks like typing, musical instrument playing, stone-carving, archery and fencing require extensive training, spanning years. Trying these sophisticated skills is often daunting, leading many to avoid them. We shy away from them due to their challenging nature, resulting in the underutilisation of our ability to practice complex coordination and merge initially foreign movements into a new, comprehensive skill set. We tend to stick to what we're familiar with and enjoy the activities we're already good at. When beginning to acquire a new skill, especially as an adult, it's natural to feel terrible initially. We often prefer to focus on the functional aspects of our physical life – improving our strength at the gym or endurance and stamina on the treadmill, which, however vital, are less demanding or challenging to our physical intelligence.

The thing is, coordination is something you need to work at, even if you're a professional mover. It's not automatic; coordination must be brought into our awareness.

My choreography, for example, is known for its intricate and demanding nature. Dancers join the company precisely because they want to test and develop their physical abilities in the context of creative play. This means that they often have to unlearn movement patterns that have become ingrained and face many complex coordination challenges. This can be exhausting; however keen they may be, the body and mind sometimes resist. Such resistance is natural; we want to fall back into what we know and take the easiest route, as this is part of our survival instinct as humans – to save energy and resources. But these needn't be mutually exclusive – it's just that you must experience the uncomfortable feeling of not knowing for a little longer than you may like. Eventually, with the right approach, any intricate coordination can be achieved, and the body and mind are inspirationally nourished in the process.

Play

Throughout childhood, we experiment with what the body can do. From infancy to around age eight, this occurs through bursts of acts, which may seem random to adult eyes, and become the foundations for functional motions once the brain begins adding motor planning to movement. We work out how to reach, then touch, then hold things. Coordination is aided greatly by this hand–tool play. These seemingly aimless motions also contain questions. What does my arm do? How does this feel? What does it weigh? How to hold? What is force? What's up? Down? Gravity? How do I? How does this? What happens if? Our flesh and our environs: a wonderful laboratory.

At around seven months old, it dawns on us that we are not part of our mother but singular. Soon, we pair vision with proprioception, matching what we feel we're doing inside with how it looks from the outside. Before we form a body image, we're pure schema.

In addition to free and structured play, we absorb motions and behaviours tacitly by noticing, watching, copying, and mirroring, testing others' actions with our bodies. Even at a young age, we're sharp – we'll only mimic moves that appear sensible, given our burgeoning world conception, and seem to fulfil their function.

Essentially, coordination results from attentive motor skill development, which enables the appropriate interactivity between speed, distance, direction, timing, joint position, and muscular tension to be transformed into a bodily action pattern and mental movement schema. This can only be accomplished through purposeful, careful work.

Soon enough, we start planning and coordinating original moves to achieve the desired results, gaining ever greater control over our bodies and behaviour. We form habits. In this way, little by little, we furnish our physical intelligence library and sculpt a unique movement style: our lifelong physical signature. Over time, the coordinations we practise become fully proprioceptive – we can do them 'without looking'.

The coordination we learn during our earliest years is relatively universal. Beyond that time, we keep honing and tuning our instrument's capabilities along available avenues, developing swathes of regularly practised skills into proprioceptive and habitual actions and a few into full-blown virtuosities.

Yet, as many of us age, we become inhibited. Our inner voice tells us that trying a new thing is too hard. Or we give it a go but give up the second we find any element frustrating. When we do this, we short-change ourselves because, whatever we might think, we all remain capable of learning complex new coordination; that extraordinary facility doesn't just disappear.

What changes is that although learning by mimicking plays a major role in childhood development, copying isn't the ideal way for adults to tackle novel coordination. By the time we mature, we absorb this sort of information in discrete, sequential stages. We learn one action at a time and then memorise them as related chunks. For coordination to become complete, the body and brain need to figure out these units and chunks and matching patterns and schemas for themselves.

It's also good to remember that the strongest indicator of poor motor performance is not reaction time or cognitive speed but our (in)ability to pay attention. This is because, at its essence, coordination results from teamwork between the muscles and conscious intention. If we persist at an activity for long enough – even (especially) when we feel as if we're at the edge of our comfort zone or when there's a voice within us screaming, 'I can't!' – we give ourselves the best chance of mastering it. This remains true whether we're tackling an extension of a familiar skill or something bracingly new.

The older we are, the longer we've resisted physical experimentation, the more discomfiting this process will be. But it is here that dance – an art based upon the execution of deviously complex coordinations – can help us.

Complex Co-ordinations

As a choreographer observing dancers as they tackle new sequences, it's crucial for me to provide them with the freedom to navigate their learning process, allowing them to understand through 'mistakes'. By enduring physical trials and errors, the body naturally adapts and learns. I refrain from excessive early corrections or attempting to adjust their movements. Instead, I allow the learning process to unfold naturally, even if it means letting inaccuracies persist momentarily until the dancers and their bodies recognise and adjust their coordination independently.

Professional dancers are willing to go through the first, second, and third iterations of a physical problem – as many as it takes for the coordination in a movement sequence to land. My role is to encourage that dancer to go through those iterations. If, after the first pass, I flood them with a lot of technical information about what they've just done or overcorrect too soon – 'No, that's not what I meant.

Look, do this!' – they won't get to that fourth iterative version, nor will they brave the unfamiliarity from the inside out. And what revelation and delight there is when the phrase that challenged our coordination settles.

It's akin to the experience of deciphering stereogram pictures – where what seems to be a flat, repeating pattern suddenly transforms into a 3D image right before your eyes. This requires effort (sometimes you need to stare at the image for a while), but the sense of achievement is unparalleled when you finally see it. For a dancer, embodying intricate moves for the first time – be it a fluid backhand, an effortless handspring, or a perfectly executed Fosse routine – is among the most gratifying sensations one can feel physically.

This is true for you, too. You learn by allowing yourself to go wrong, giving yourself the time and space to make mistakes, and ensuring you don't overcomplicate what is already a challenging endeavour.

An initial step to mastering any action series 'without looking' is understanding how to 'look' at it beforehand: how to break it into composite parts. Try to attend to only part of the coordination at a time. Tackling complex coordination as a whole is a fool's errand because it contradicts how we learn by sequentially chunking and layering elements. It is in our nature to figure out individual components and to find some anchors, which you can then build iteratively.

This practical choreographic framework is for parsing any motion series by unpacking its *what*, *where*, *when*, and *how*. The *what* is the part of the body that moves, the *where* is its direction through space, the *when* designates timing, and the *how* is the applicable amount and type of effort and force. A simple kick is a good example of how this might be applied in practice.

Pull one up from memory. No matter if it's football, kung-fu, hip-hop, capoeira, or can-can. Review the mental 'movement-image' (no harm in scoping it out online if a refresher helps). Reconsider it moment by moment, part by part.

What is the extending limb actually doing? Is the knee straight or bent as it rises? Is the foot flexed or pointed? What is the standing leg doing? How ought hip, knee and ankle align? What is the spine up to? Which way is each hip facing at the kick's highest point?

Now, think about *where* — about the movement's direction through space. What is the raised limb aiming for? Forty-five degrees above the hip? Straight out of it? Thirty degrees towards Earth?

Consider *when*, its timing: which muscle groups move when in each leg? How does the motion unfurl: sloooooooow, quick, quick? Quick, slooow, quick, quick, slooooooow?

How much force demands applying, and in what way? Are you pushing down to go up? *How* much energy needs expanding, and at which stage? Is the movement a coil sprung? Hop-float? Press-punch?

★ ★ ★

The longer you play, the more practical details will emerge. This framework is helpful for identifying the components of movement. However, as we learn these components, we tend to develop a preference or natural bias regarding the *what*, *when*, *where*, and *how* of our actions. We tend to gravitate toward the concept we are most comfortable with.

Whether professional dancers or more sedate homebodies, we all have a wealth of tools in our physical thinking portfolio that lead to this bias: perhaps a feel for rhythm, hereditary (or hard-won) power or flexibility, an affinity for capturing the macro-line of a lengthy sequence, or for rebuilding it iteratively, micro-gesture by gesture; acute visuospatial awareness; a photographic or kinetically geared memory. And we will do well to play to our strengths as we commit to this physical learning task. For example, some may be drawn to *where* – to a movement's route through space; some to *when* – rhythm through time; others to *how* – to dynamic and force. Any of these can serve as an 'in'.

In Company Wayne McGregor, some dancers are adept at replicating shapes but struggle to convey a sense of momentum. They focus more on *what* they are doing rather than on *how* they are doing it. When learning a new dance phrase, some dancers prioritise replicating the shapes over capturing the movement's dynamic essence. On the other hand, some dancers excel at understanding a phrase's speed (*when*) but need help to specify which body parts initiate the movement (*what*).

I encourage the dancers to start learning a dance phrase by noticing what attracts their attention first (while, over time, evolving their skills). I want them to use their natural preferences to begin the ingesting process, as people are usually more comfortable relating to things they have done well in the past. Then, I ask them to consider the other aspects of the framework to enhance their initial understanding of the phrase. It is at this stage that progress is most noticeable. The framework helps the dancers recognise their 'looking' habits and gives them a range of new options to embed the coordination in their bodies.

Interestingly, however long a dancer works on trying to accomplish complex coordination, sometimes they need to sleep on it. The next day, it has usually settled, and they are ready for a new challenge. Memory and sleep have a closer connection than you might think. While you sleep, your brain moves important information into long-term memory. You store the physical inputs and experiences to build on them and do it again. And by contrast, if you're tired before you even begin learning, it inhibits that process of embodiment. This means that if, while you're learning a new skill, you make sure you get enough good-quality sleep, you'll accomplish that process much more effectively than if you stay up all night and risk that learning vanishing. The more sleep you get, the more effectively that new skill will become more embodied. With that novel physical memory laid down, parsed, and streamlined, you can build on it the following day. I'm pretty deliberate about offloading memory before sleep – it's a little like doing a mental checklist. When I was younger and learning longer choreography sequences, I'd try to focus on the phrase I would need for the next day before I slept. Although I focused on a dance phrase, it's an approach that would work for any coordination. It's like the sorts of visualisations prescribed in books written for executives. The sort that might suggest that before going to a job interview, it would be useful to visualise the route you'll take to get there.

Critically, this system of parsing and learning needs to be utilised holistically to search for the relationships between *what*, *when*, *where* and *how* rather than staying with them individually. It is in the mixing and combination of these distinct parts; it is through the transitions between the elements analysed, that we can build back the logic of the dance phrases into a more coherent whole. This takes us from watching and breaking down the dance into a series of discrete bits to a chain of motion. To continuous flow. All dance is about states of flow – the transition between elements, their interplay as vital as the elements themselves.

Connected Parts

Our conception of the body can very often be of a collection of discrete parts, each operating individually. But I think of it as this beautiful machine in which everything is linked. It's not an archipelago of organs operating independently – it's a system. Every part of your body is connected to the rest of your body – even if sometimes these connections are not immediately apparent.

Think of how your nervous system (the brain, spinal cord and nerves) receives and processes information from the senses and sends instructions to the muscles and organs. Or how the glands comprising your endocrine system produce and secrete hormones, which then travel through the bloodstream to all your organs and tissues, regulating functions such as metabolism, growth, reproduction and stress response.

I like to use the concept of chaining when dancing (or learning complex coordination). To help others understand this, it's a way of (re)creating that feeling of flow essential to dancing bodies.

The idea of chaining is both practical and imaginative – it's about how the body is in connection with itself. Think about the 1970s and 1980s dance style of body-popping, where you could follow a sequence of muscle movements that would begin in one limb of the dancer and then travel around their whole body. The movement could start in my heel and flow up to my knee and then hip, through the side of my back to my other arm. It's a wavelike strut that moves freely, completely unimpeded. Our pleasure in executing the move or watching others performing comes from witnessing this wave that travels from one place to another in a seamless stream of motion. Some of the best body-poppers look boneless as they show us the detailed pathway through the body of a shape that started in the hand and snaked to the dancer's head and back again. As you imagine this motion, you see that movements are all connected; they are not static movie keyframes or animation stills. They are not isolated body parts (*what*), moving in arbitrary directions (*where*), at random speeds (*when*), with no quality (*how*). They are instead an integrated system – a chain of movement events slipping and sequencing together fluidly. Each relies on the previous and subsequent moves to create an organic transition between them, generating flow. This body system is working collaboratively and thinking holistically.

Or, for another example, let's look at a dive by Olympian Tom Daley through this lens.

Imagine Tom is drawing with his body, and instead of his limbs being limbs, they are paint brushes. Notice the marks he makes (the brushstrokes) while focussing on one aspect of his body. It's a kind of action painting. Start with his hands (*what*): Tom's hands reach outwards to the top of his head in preparation for the dive. We can 'see' the two wide arcs he has painted on either side of himself. As Tom bends to jump off the board (slow/down fast/up – *when*), we see two parallel brush strokes: the first on his bend, which is short, and the second on his jump, which is long (*where*). In the air, Tom tucks and rotates thrice while falling towards the water. Our drawing sees the hands come close to his knees and grip tight (*how*) while we paint three roller-coaster loops descending. Tom stretches, his hands back above his head, his body close to the water and splash. The brushstrokes lengthen, narrow to two parallel lines, and both are drawn quickly down.

If we now replay this action painting in our mind or even trace the motion with our hand in the air, we can recreate Tom's dive as a drawing: arcs, straight lines, loops and curves. We can imagine the geometry and shapes made by Tom's hands (even without Tom in the image); we can remember the speed of the moves and when the speed changed, and we have embodied a sense of the motion as a whole. By following the complete pathway, by chaining and connecting the dots, we have a very different overall understanding of the action than if we had broken it down in any other way – a complementary way of seeing and a fresh take on the complex coordination.

Chaining, then, gives you a wonderful sense of a phrase's overall motion, its transitions, and its beginning, middle, and end – the complete movement, the 'thought', as it were.

This concept of chaining is critical to both watching and learning dance and to dancing itself. When dancing or performing any physical task, the seamless interconnectedness of our actions leads to flow. When the flow is interrupted, we experience awkward blockages and potentially fractured motion in the body. These blockages may expose the awkwardnesses in the phrase you are learning, the parts for which you have not yet found coordination, or they may indicate a block in your body.

A blockage anywhere in our body's chain traverses our physicality. (This is partly because muscles, tendons, ligaments and joints are encased and connected by collagenous tissue known as fascia, a key component of our embodied force transmission system.)

It gets even more complicated when you realise that the part that needs to be fixed is not necessarily the part that needs attention. A dancer might jump for years, landing slightly on the outside of their foot. But they won't necessarily have a foot injury; it might lead to incredible backache. Perhaps your shoulder is painful. The pain might not be generated by or in the shoulder. If you were to chase it down, you'll likely find it comes from somewhere else in your body – maybe a tightness in your hip that you've created because you're never stretching your back. No amount of shoulder work you do will release the shoulder because you're not concentrating on the part in the chain that's not working.

Sometimes, the impact of these connections can be even harder to trace. If, for example, someone experiences chronic stress, encouraging the release of cortisol (stress hormones), this can affect multiple systems in the body. Elevated cortisol levels can disrupt the balance of our endocrine system, leading to potential alterations in our metabolism, generating issues with our immune suppression, and prompting disrupted sleep patterns. These changes could then potentially impact other systems, such as cardiovascular health, digestive function, and mental well-being.

This relationship between an injury or discomfort in one part of the body and an expression of that injury or discomfort elsewhere is the basis of those holistic approaches to health and well-being that consider the body an interconnected whole. These practices have the capacity to teach us much about how to maintain the integrity and health of our bodies. Mindfulness techniques, such as meditation, deep breathing exercises, and body scans, can help you tune into your body's signals and identify areas of imbalance or discomfort. Stress-management strategies, which can be as simple as spending more time in nature, can reduce the impact of stress on your physical well-being.

WE ARE CO-ORDINATION

Reflexology will tell you that there's a point in your foot which can tell what's going on in your liver and that we're all parts of these energetic systems that are talking all the time.

But we tend not to think about it.

When one considers the ways in which the body is connected and tries to put this knowledge into practice, it can have a significant, positive impact on our health and well-being. It can also be a foundation for learning new physical schema and fostering creativity.

Achieving complete coordination means we've acquired a new skill and another physical possibility for our instrument to play with. And that our attention can now be diverted to the next goal. In dance, this might mean fine-tuning the just-learned phrase's emotional expressivity. If our choice coordination involved playing a new piece on an actual piano – a virtuosity of the hands – we may now disappear into that music. If touch-typing is our freshly conquered task, we're entirely free to flow into words. Once we've practised any activity long enough to access flow through it – to dissolve into it – it feels effortless.

6
We are Chemistry

What are two of the most stressful regular life scenarios you can imagine? First dates must rank pretty highly. Think back to your most significant one. Can you recall the hours leading up to it? How alive you felt as that electric fizz coursed through your system, those relentless flips in mood and attitude? From peak child-at-Christmas elation – what if this is it? What if they're the one? What if we fall in love? – to gut-tanking bouts of anxiety – what if they stand me up? What if I've nothing to say? Why am I wearing this? Wait, why am I sweating?

Then there is public speaking. All those words, facts, and concepts to remember. A blur of expectant faces. That urge to flee, to just back out. Wild pitch shifts, from 'I've got this. Let's go!' to 'Not a chance. They'll see right through me: help!' Incessant stomach twists, jittery breathing, damp palms, clammy pits, and dry mouth.

How about imagining that you – not a scientist by trade – were scheduled to give a talk at the Royal Institution, one of Britain's most illustrious scientific organisations?

Want the ultimate nightmare? How about both those things – speaking at the RI *and* a first date – taking place on the same Valentine's Day, one straight after the other? That was the situation I found myself in some years ago. I was a fraught mess. On my journey into London, I came close to bashing the tube station's fire alarm and running for the hills. An internal alarm was already going off – my breathing was frantic. I paused, paralysed, smack in the middle of Piccadilly.

I knew there was no way I'd do myself (let alone the Royal Institution) justice if I attempted to speak from this panicky state. So, as I walked past the Ritz, I shut out the surrounding bustle, hushed that angsty mental commotion, and committed to recentering myself.

I drew my attention inwards, focusing solely on modulating my breathing, feeling my pulse slow as I connected to my body's still point. Once I was calm in body and psyche – relaxed, aware, and present in my surroundings – I strode on, reached the Royal Institution, entered the auditorium and positioned myself at the lectern. The talk went well, and the date did too (Antoine and I have now been together for twenty years), largely thanks to that physically intelligent, attentional and energetic reset.

Have you ever noticed how all anticipation, whether of something fabulous, such as a holiday, or less fun, like an exam, can create the same feeling in your body? All excitable restlessness and wired energy. Recall how much effort it takes to keep it under wraps – to rein yourself in – as the sensation of something significant just ahead, ecstasy or disaster, builds.

These intense states can feel so alike because, essentially, they are. Whether preparing for a positive or negative event, our mind and body boost us to the place that will likely serve us best by using the same hormone. Hormones like cortisol, known as the stress hormone, affect how we respond to stress and influence our emotional awareness; oxytocin, often called the 'bonding hormone', facilitates social connections and emotional awareness, making us more sensitive to our internal feelings; and adrenaline, released during stress, impacts how we perceive our body's physical state. These chemicals work together to help us understand and interpret our internal bodily sensations, which is crucial for managing our emotions and our overall well-being.

As adrenaline circulates, for example, our heart beats faster, and our breathing pace accelerates. Extra oxygen floods the brain as glucose flows into our bloodstream to help fuel our next moves. In a flash, we're hyper-alert – attentionally primed – and fully energised: raring to go! While this state of high energetic arousal might feel volatile, it's the ideal place to be with a test to ace, a theatre crowd to entertain, or a single gorgeous soul to impress.

Some people genuinely relish, even thrive in, these amped states. You might be the type who loves exams and feels vitalised by pressure-cooker scenarios. This isn't the same as getting off on risk, though adrenaline junkies might be found in this clan; it's just that we all operate differently in high-stress situations. Many can't hit their stride without duress, whereas others become discomfited, sensing themselves shutting down.

Energetic Arousal

Over a decade ago, I noticed these kinds of energetic peaks and troughs in the studio while working with my company. Some artists were swift to attain creative flow: they would arrive enlivened and sparky, then soar upon that collective pressure to deliver expressive ideas quickly. But other dancers, who were just as technically skilled, physically fit, and artistically gifted, needed time to acclimatize and get into their zone. And then, even once motoring, they were readily thrown by too much attention, let alone intensity, from me. Yet the naturally adrenalised dancers had limits, too. They would eventually begin to flag, then hit that proverbial wall.

Keen to adapt my approach to these sessions to better engage, inspire, and support the entire company, I wanted to investigate how the dancers' bodies responded to stress while creating together. So, I approached the technologically progressive London design duo Studio XO with an idea for a research-based dance and design collaboration.

This was 2012, before Fitbit debuted their first heart rate-tracking wristband, when biometric watches (which measure vital statistics) were almost unheard of. But as Studio XO specialised in translating emotive experiences, such as first dates or elite-level dance-making, into data, they'd already developed arousal-reading wristbands. They and their ground-breaking emotional technology joined the dancers and me in the studio as we choreographed Atomos. Every dancer wore a wristband that tracked their pulse and monitored how much electrical activity was in their skin. This indicated how aroused (energetically and emotionally) they were at any given time.

The data we amassed allowed us to 'read' each dancer's energetic fluctuations and varying engagement levels as we progressed through weeks of in-studio movement generation. As expected, these internal chemistry charts highlighted the extreme physical demands of devoting one's full attention to creating under pressure at superhuman velocity. And the dancers' stats jumped higher still during periods when my focus was directed at them personally. Studio XO then transformed this data into soft sculptural objects that the dancers worked with creatively in the studio. Data collected from those interventions was used to design patterns for the company's skin-like Atomos costumes.

The information harvested by the biometric bands was a revelation. Even though dancers are usually acutely bodily aware and able to discern and manipulate many of their own internal signals, they had registered little about their pulse rate, levels of engagement and emotional stress while working creatively. They had never been able to access that information as part of their physical intelligence toolbox, let alone figure out how to manage or focus techniques to reset themselves quickly if under pressure.

After reflecting on those results, I began dividing my attention more equally among everybody present at every company rehearsal. This allowed for periods of intensity when the dancers were most energetically aroused, balanced with periods where they were not the point of concentration. Their attentional resources were 'managed' more carefully. I also started capping my choreographic sessions, including those with the Royal Ballet, at one hour per dancer or group, to make it easier for us all to sustain energy throughout each day, across each six-to-ten-week dance-creation period, and beyond into multiple years of collaborative artmaking.

Energetic Baseline

It's useful to remember that everyone has a kind of energetic baseline. It would help if you considered it as your neutral state. It's how you feel mentally, physically and emotionally when all your needs are sated, nothing's ailing or niggling, and you're pretty sure you're not about to be plunged into any drama. It's the state you're at ease in — altogether level, effortlessly yourself. Everyone's energetic baseline will be different because we're all unique. Some of us are relentless balls of energy; others exist in a far calmer, contemplative place. And given the daily pressures we contend with (and the stimulants, such as caffeine, we take to get by), it may have been a while since you rested in neutral. So, it might take a moment to recover that state. Our ability to identify how energetically aroused (or not) we are is not just something we can use to prepare ourselves for something taxing, such as a daunting speech. It can also help improve our ability to communicate. Because as important as it is to be able to modulate our physical and mental state to reach a place that is more conducive to our needs, it's also beneficial to have the capacity to attend closely to what other people are saying and doing (rather than drifting into a reverie or scrolling on a phone). To clearly and sensitively convey our thoughts, intentions and emotions to others – to give ourselves the most potent shot at being heard and understood – we need to prepare ourselves to engage in the first place. The outcome will not be pretty if we enter a draining work meeting already depleted. If we barge into a cosy dinner with a newly broken-hearted friend while boiling over with agitation, that mismatch will muddy conversation and hinder genuine connection. Think of a close friend. What does that friend seem like when they are at their energetic baseline, at ease with themselves, their situation, and their environment? How does this make them move and act? Are they relaxed or sparky? Are their movements languorous or bouncy? Do they speak at a snail's pace or gabble away at 100mph?

Like much of the population, their neutral probably rests somewhere in between. And given that you are best mates, it's just as likely that your neutral energetic baselines are complementary: it's effortless to be in their company when you're both at your baselines.

Think now about how your friend 'reads' at those high and low arousal extremes. How differently do they present in those oppositional states? How expressive is their face? Does their complexion change? Do they appear to grow in stature or kind of disappear? Are they jittery or jarringly static? Do they display many of their habitual moves? Or are signature gestures deployed at a disturbing pace? Do they appear sharp and focused, scattered and wild, plain zonked out? Are they open to touch contact in either state? Can they attend beyond themselves? And do you feel that they are still present with you?

Before we even open our mouths, our energetic state projects our positivity or negativity ahead of us, silently impacting the interactions that follow. If we meet someone when we're in a heightened or depressed state, that person will detect that we seem 'off'. They will not know if we've had too little sleep or too much sugar. Still, they will register that we're distant or defensive, overexcited and inattentive, and likely imagine it reflects how we feel about them, that relationship, that project. This perception will alter their responses and communication style accordingly. That might sound drastic, but we all do it to a greater or lesser extent all the time.

Using our body and breath to shift our energy into neutral or anywhere else along the arousal spectrum can make a massive difference. We must cultivate the habit of making sufficient time to attain a suitable state of preparedness and re-centre ourselves ahead of engagements. Before you meet others, discern what kind of energetic signals you're emitting. Determine whether the encounter ahead needs a quick blood-sugar-stabilising snack or a snatched ten-minute nap.

Depending on individual requirements, you can manipulate your energy levels hours or even days before major events. For dancers, this might mean banking extra sleep and additional preventive bodywork before an international performance tour. Marathon runners might wind down training runs and feast on carbohydrates to stockpile personal-best-beating energy.

One can even predict, plan for, and finesse one's state of preparedness for major life events (such as a new baby or a big operation) and/or essential work commitments (for example, an intensive film shoot or a draining election campaign) weeks, months, even years ahead.

Let's turn inwards for a moment.

Shifting Energetic States

Lie down. Place your hands gently on your lower abdomen.
Breathe solely through your nose.
Inhale deeply, to the bottom of your lungs and beyond, feel your tummy rise.
Once full of air, exhale completely. Don't force it out; just let that air escape.
Repeat for a couple of minutes. Perhaps you'll soon be able to tune into your pulse. Listen.
Begin to inhale to the count of four, pause, and then exhale to the count of eight.
Repeat for two more minutes. Note whether, even if only briefly, your consciousness rests solely in the body.
Stop counting. Keep breathing slowly, entirely – without any pauses: circular breath.
Become aware of how, by inhaling, you're bringing life energy and oxygen into your body.
Once you've delivered that energy and life force, exhale.
As well as providing a refreshing reset and a shift in our energetic arousal, those few minutes of focused breathing will have highlighted that the body itself is a container of space. When we breathe in, we fill it, replenishing ourselves before emptying it once again. The deeper we breathe, the more profound the effect on body and mind. When we

breathe through our nose, we optimise both the volume of air we inhale and its cleanliness (because our nasal hairs filter it).

We've evolved to breathe down into our diaphragm rather than with shallow chest-only inhales—puffing out and deflating our tummies like babies. The problem is that, societally, we're stressed out. The more tension we hold, the slighter and faster our breath – further upping our arousal, which the body detects and relays back to the mind as a warning sign, increasing mental anxiety.

Body image can interfere here, too. Sucking in our stomachs to look thinner, donning restrictive clothing, constraining our diaphragms, and forcing breathing to remain high and shallow can quickly become a detrimental embodied habit.

Luckily, because breathing is connected to every other embodied faculty, it doesn't take long to calm it all down – it's really easy to reset ourselves. Once you're comfortable with circular breathing (or if you're already a committed yogi or Wim Hof disciple), you can start attending to how much internal communication you can detect while practising. Connect to your heartbeat, sense your blood circulating – listen out for awakening bodily consciousness.

The benefits of doing this go way beyond steadiness of breath, lower stress, and a quieter mind. Just a few minutes a day of parasympathetic breathing (breathing with elongated exhales) has been proven to lower blood pressure, increase immunity and aid digestion, among other benefits. This is because the parasympathetic part of our nervous system, which we 'switch on' by breathing slower, is the body's rest, digest and heal mode: the zone we segue into when safe, sated, and relaxed. And you can opt to drop into that state whenever you want!

Imagining that your body is a container of space helps open avenues of awareness. While our lungs and other body parts are frequently depicted (and thus imagined) as flat-plan illustrations, we are in fact voluminous three-dimensional vessels! Breath can be placed wherever we choose to 'send' it. Just as we can create visualisations to aid us in interpreting the space around our body, we can devise personal imagery to guide us in directing breath around our insides. For example, we might interpret incoming oxygen as a ray of light and shine it at the back of our lungs to sense their true capaciousness or beam it towards blocks in our bodily chain as we focus on how good it feels to have all that stiffness melt away.

When we 'switch' the body to its rest-and-digest mode using our breath, we note its profound effect on our mental and emotional state. To balance ourselves out, we can attend to how we reverse this process to purposefully activate and energise ourselves whenever an energetic level-up is required.

When I discussed how stressors (such as public speaking and first dates) tip us into anxious and/or excitable states, I was describing what it's like to reside in another nervous system mode. This zone of heightened arousal is commonly known as our 'fight-or-flight' state.

But being in fight-or-flight isn't always negative, even when we've nothing to attack. Sometimes, we need to shift ourselves into that place to manipulate our internal chemistry and charge ourselves up as much as possible. Perhaps to access an electrifying performance persona ahead of opening night, generate pre-birthday party energy, or even sprint-finish a book chapter. Just as we can calm our entire being via breath alone, we can gift ourselves an energetic level-up anytime we like.

Pre-show dancers often try to boost their energy. This might involve running laps around the stage, skipping in place, or practising their most complex moves. Many prepare through yoga sequences and breathwork. Jess (a veteran dancer with my company) uses the 'Breath of Fire' technique to stimulate her nervous system by increasing her body temperature, heart rate, and blood flow. This leaves her feeling dynamic and more prepared to tackle the problematic performative tasks ahead.

Breath of Fire is a distinctive breathing exercise practised in kundalini yoga. It involves rapidly expelling air from the lungs using the diaphragm, the primary respiratory muscle located near the lower ribs, and the intercostal muscles. While many yogic breathwork exercises emphasise prolonged deep breathing for relaxation, the Breath of Fire technique is designed to energise and invigorate the body.

Sitting comfortably on the floor, your spine should feel like a straight line from your neck to your tailbone. To centre your body, focus on the present moment inwardly by consciously thinking about your solar plexus (the cluster of nerves below your sternum).

Once you feel ready, pick up the pace of your breathing, with inhales and exhales of equal length through the mouth. If it's your first time practising Breath of Fire, aim for each breath cycle (from inhale to exhale) to last about one second in length; it should sound similar to a dog's panting. As you breathe, channel the air through your stomach rather than your chest. Place a hand on your navel point to feel your stomach expanding and contracting like a bellow with each breath.

After a comfortable rhythm, close your mouth and begin channelling breath through your nostrils. You are now performing Breath of Fire. If you're a beginner, continue this practice for up to fifteen seconds before returning to a more natural breathing pattern. As your lung capacity increases, you may find that you can safely continue fire breathing for up to a minute at a time.

Breathing yourself in and out of different energetic states means attending to your body through its internally geared perceptive systems. Think back to our early section on how we attend to and perceive the world using senses such as vision and hearing. These externally directed senses are crucial to how we interpret our environment. But as far as body and mind go, they only form part of the perceptive picture.

Somatic Markers

An insect zooms around the room you're sitting in. How might our interior perceptive faculties 'read' that buzzing noise? Recall the last time you were being 'stalked' by a random insect. Did it make you edgy? Perhaps it induced a searing flashback of a childhood sting? At the same time that our ears and mind 'discuss' the noise that this bug is making, our body generates an internal perception of the insect and a feeling response.

If you were stung horribly in the past, you might encounter a spectre of that pain and anxiety, even if the buzzing sounds more like a fly than a wasp. This disquieting sensation might lead to the diversion of your conscious attention as you try to determine what sort of animal has invaded your space.

Visualising (imagining) any intense experience, especially while delving into stressful memories, will likely cause the mind and body to generate a shadow play of a similar state: it will increase your arousal levels. This process is a facet of how we're able to communicate our experiences, empathise with other people, and lay down memories. Although we mostly think of our recollections as visual or audio snapshots, an imprint of how we *felt* during any given moment, such as those before a hot date or during a hornet duel, will be stored away, too. This is why, as we reflect on past tribulations – no matter if we pulled through these encounters unscathed – we may feel tense all over again. These embodied memory imprints are a kind of somatic marker, a body's way of storing feeling in and as an 'image'. In dance, we call these 'kinaesthetic images', and alongside visual and acoustic images, we shift between them all the time in our dance training, performance practice, and creative processes. More of that later.

Our perception is active: it helps to guide our attention to potential harms (such as stinging insects) as well as spotlighting chances to sate our needs (snack time). I've already discussed how our minds and bodies continually reinterpret our environment, providing us with action boundaries that inform how we move through space. By combining those ideas with these ones about memory, we should now be able to understand how what we desire or dread directs our attention as we go about our day and, in turn, alters how we perceive the places we encounter.

Remember these examples from Chapter 3 – if we're hungry, we'll be primed to notice a cafe. Or, if wiped out, we'll be drawn to locations quiet enough to rest in. More drastically, if feeling threatened, wherever we find ourselves, we'll become more aware of viable escape routes. All the while, primed by our needs, we're mentally 're-drawing' our surroundings. In this way, spaces, even those that are technically neutral, are not neutrally rendered by our body, mind, and psyche. Much like other people, they are, at a base level, attractive or repellent. Even if we're walking along the street, preoccupied by a work crisis, we're always attending and thinking physically – drawing upon our perceptive powers to appraise how appealing or antagonising every feature of our locale is to our unique state at that precise moment.

Interoception

Fundamentally, these extraordinary internal systems are primarily occupied with keeping us alive. Every body contains integrated networks of such systems, which detect and signal whether we're hungry or thirsty, exhausted or revived, anxious or calm, hot or cold, coming down with the flu or in rude health.

Whether we are wide awake or fast asleep, these critical faculties alert us to base needs – breathe, eat, drink, sleep – as well as to pleasure and pain or discomfort and ease. They pinpoint pathogens, heal injuries, and evaluate immunity. They minimise or amplify our feelings. They detect our levels of vital nutrients and toxins. They interpret our overall energetic state. In this fashion, in tandem with our minds, these systems strive to ensure we attain and maintain internal equilibrium (homeostasis) by adjusting our body state to meet our *predicted* energetic needs (allostasis). This collaboration is a remarkable example of our physical intelligence – where representations of our current physiological status are modulated to suit our future physiological status needs best. Our nervous system anticipates, senses and integrates our body's present state signals to enable, support and nurture its next.

Beneath the highs and lows of experience – victories and disasters, desires met, and dreams forsaken – we are also always just this: a body and brain figuring out, moment by moment, how to stay alive on this planet for a bit longer. I find this idea startling and beautiful. A significant part of this effort to keep alive are these internal systems guiding our attention to opportunities to fulfil needs and elude harm.

Together, these networked internal perception systems are termed 'interoception', which was defined in a 2018 paper by Dr Wolf Mehling and his colleagues as 'the overall process of how the (total) nervous system senses, interprets and integrates signals originating from within the body, providing a moment-by-moment mapping of the body's internal landscape across conscious and unconscious levels'.

There remains much we do not know about ourselves. What is consciousness? Which possesses more power over pain, the body or the brain? How does our 'gut brain' (that is, the enteric nervous system) influence our thoughts, actions, and feelings? What role does the proposed 'heart brain' play in our intuition, decision-making, and emotional lives? We cannot definitively ascertain whether inflammation is a cause or a consequence of depression or how it interacts with hormone levels during menstruation. However, there is sufficient ballast in the burgeoning interoceptive research field for us to now consider how it affects our well-being and behaviour.

Interoception: Feeling + Emotion

Earlier, while imagining ourselves preparing for exams and dates, we considered how varying levels of tension and excitement – energetic arousal – influence us and our actions. These embodied feelings, known as affective states, arise from interoceptive processes. Emotions, too, stem from our awareness of internal bodily sensations.

An affective state refers to an individual's overall emotional condition or mood at a specific moment. Such a state is typically more generalized and tends to last longer than a brief, specific emotion, which is a response to particular stimuli or situations. Emotions are often short-lived and can arise as reactions to events or thoughts, usually directed at or triggered by an object. This object can be a person, animal, environment, fact, artistic creation, or an abstract idea. Although we often consciously interpret our feelings as emotions – even though they are related but distinct concepts – and may sometimes not even notice them, they are always present within us. We typically do not notice changes in our blood pressure or heart rate fluctuations as long as they remain within healthy limits. (In fact, heart rate variability can indicate good cardiovascular fitness.) However, we may experience other physical signs of arousal, such as clammy skin or rapid breathing.

The way we interpret arousal – whether we perceive it as pleasant (like on a rollercoaster ride) or unpleasant (like during an exam) – depends on several factors. These include our overall mental and physical state, what we focus our attention on, relevant past experiences stored in our memory, and the context of the situation we are in. The overall positive or negative inclination applied to our arousal is referred to as its *valence*.

In her insightful book, *How Emotions Are Made: The Secret Life of the Brain*, Lisa Feldman Barrett suggests that the best way to understand our affective and emotional states is through these two key dimensions: arousal and valence. Low arousal combined with negative valence indicates a decline in energy, often due to overexertion. Conversely, low arousal with positive valence suggests that we might be relaxing on vacation or enjoying a moment of mindfulness. High arousal with negative valence can result from feelings of fear – like when facing an approaching hornet – or anger, such as a response to current political events. In contrast, high arousal with positive valence reflects excitement, such as the thrill of a promising date or the exhilaration that comes after dancing, fuelled by endorphins. Overall, learning to access and interpret affective and emotional states – in a distanced, curiosity-led manner – is a phenomenal manifestation of physical intelligence and a game-changer regarding behavioural control. How are you feeling right now? Are you relatively calm and contemplative? Slightly fraught, agitated? Review your body state – then explore what might be behind this mood from an interoceptive point of view.

> Have you eaten enough today?
> Did you sleep all right last night?
> Is there a stressful appointment coming up tonight?
> Did you have to skip your daily run?
> Is an aching back or irritated bowel grouching beneath your conscious awareness?
> Are you coming down with something?
> Or perhaps it's just too cold wherever you're reading this book?

All these interoceptive measures – 'reads' of internal chemical states – will always determine how you feel. It's perfectly plausible to be in a vile mood, convinced the world's out to destroy you, then trace it back and realise the root cause is crashing blood sugar, an imminent cold, or chronic 'physical boredom' – that icky trapped feeling that comes with under-exertion.

As infants, we get audibly fretful when we're hungry. To a lesser degree, this happens till the day we die, no matter how mature and supposedly shut off from our bodies we are. The same goes for the restlessness that partners with borderline exhaustion. But sometimes, a mood is passing affect – there's truth in that cliché 'moods are like weather'. So, next time you feel dreadful, and there's no external cause for upset, check if the grimness might be the body clamouring for attention before lashing out at a spouse or cashier or having an agonized breakdown to a mate. You might just need to eat, drink, move, or sleep.

Towards Interoceptive Awareness

Discovering how emotions, physical (and attentional) states interrelate is an endlessly fascinating process – and one we each have the capacity to tune into. While we commonly mute signals as we develop emotional control – much as how we stop attending to our proprioception and body schema – this acutely personal intelligence is accessible once we (re)learn how to 'listen'.

For dancers, actors, and other performers and artists, accruing this sort of insight and then deploying it to express emotion through the body is a professional necessity. It is little wonder that in multiple studies investigating emotional intelligence, those who spend the most time 'in their bodies' as they convey emotional experiences tend to prove hugely skilled at both. This ability to accurately perceive interoceptive signals – for example, to assess heart rate and core temperature – is known as interoceptive sensitivity. This is where elite dancers and others fluent in bodily consciousness excel.

Beyond this sensitivity, having the physical intelligence to 'read', 'translate', and manipulate personal interoceptive measures generates a holistic 'interoceptive awareness' and becomes a veritable superpower!

Radical Self-care

My superpower is that I can sleep anywhere, and I might find a moment to nap now – a twenty-minute power boost to ready myself for the following commitment. The energy I need for the evening performances, meetings, donor dinners and events that will inevitably surround me with many people are totally different from my daily daytime energy requirement. For me, the social and the professional are in a river of flux. Often, it is the evening phase that I struggle with most, the desire to listen, engage, interact, and enjoy vying with the need to rest, calm, and recharge. The outward conflicts with the inward self. And it is here where my energetic baseline needs strategy, where my state of preparedness needs to be re-enlivened.

I prioritise my all-around health through a radical self-care routine focusing on mindfulness, exercise, breaks, vacations, hobbies, and eating habits, and I factor them into my daily agenda in the same way I schedule rehearsal time. This measure alone has made a hugely positive contribution to regulating energetic baseline and keeping in balance the ever-fluctuating rhythms of a day, a week, a month, and a year.

This steadfast and sustained commitment to body and mind maintenance, in whatever form, is always a net gain. For me, it's the daily infra-red sauna, shiatsu, the countryside walks, travelling to unusual destinations and constant experimentation with veganism, paleo, juice detox, Ayurvedic kitchen and sugar fasts – at the very least, nutritious food (I know that no one can perform on a diet of McDonald's, Diet Coke and cigarettes, although this knowledge is not as universally shared in the dance world as one might expect!). For you, it might be the steam room, cranial osteopathy, swimming, a four-day work week, 16/8 fasting and social time with mates that anchors you.

Radical self-care invites you to seek out the right individual work–life balance, selecting and editing your friend and support group to weed out the energy suckers versus the energy creators, setting boundaries, learning to say no even to the most brilliant people and projects precisely because you know you are at capacity, and cultivating a hard-nosed ability to ditch your concern of disappointing others – stopping the people-pleasing alone will aid you inexorably in regulating your energetic resources.

The work you do every day for yourself is not a selfish act – it's a selflessly necessary one. It's a life-long practice of keeping body and mind healthy, agile, balanced, and coordinated, articulate and sophisticated, and fluent by identifying and easing stresses and strains, re-aligning ourselves anew, working through emotional and psychic blocks, solo or with assistance, and easing tensions, anxieties, and fears so that we retain that baseline state of preparedness.

From here, even if we find ourselves in the middle of a disaster, because we've prepared extensively, we find deep reserves of energy to draw upon and the self-trust and efficacy to improvise as we navigate onwards.

But radical self-care also demands that you acknowledge, respect, and act when you have crossed the boundaries you set yourself – when life is not playing ball and you cannot find equilibrium.

If the conditions – physical or mental – are not available to us on any given day or over long stretches, especially if our role entails physical risk, as is often the case with dancers and athletes, it's wiser to step back and reset instead of struggling on, haemorrhaging confidence, and potentially suffering an injury. Sometimes, our energetic well can run dry.

Famously, GOAT (greatest of all time) gymnast Simone Biles flexed this psychic muscle at the Tokyo Olympics when, due to an inability to access the mental state required to feel 100 per cent prepared, focused and safe, she withdrew from the team competition rather than risk an accident. Her decision to *not* do – to *not* move – is a perfect example of physical intelligence. And look how that turned out for her in Paris 2024!

On a day-to-day level, a personal fear-spiking and energy-saving reset, whether mid-creative, athletic or any other process, may only demand a few minutes or hours. For example, as we stretch out a strain, exhale out embodied stress, journal through psychic knots or nasty interactions, modify our location, negotiate safer conditions, or take a nap. Or it may require a quiet day off (maybe even a year) to reassess the next best step and regain equilibrium. Developing a sense of precisely what we can work with or cope with while still attaining energetic flow takes years, longer if you are the sort who forbids themselves 'quitting'. Neither will be persistent on a continually depleted reserve. The only person who can tell you when you're on the brink of decline, the lowest energetic you – is you.

This practice of resetting needs to become habitual. At the beginning and end of each day, reset your mind, body, and space. This clears and instils an abiding sensation of calm and control, buoying your state of preparedness and re-focusing you on your goals (even if your last session was a horror show).

Once this habit is embedded, it's plausible to greet every day with a 'day one mentality', cleansed of the tribulations of its prior, revived, ready to begin anew as the least cynical and freest you!

Online Interoception

In addition to improving self-understanding, increasing emotional intelligence, and helping us navigate relationships, there's another reason it's worth building interoceptive awareness skills. As we just discussed, our internal chemistry – our levels of hormones such as adrenaline, cortisol, dopamine and oxytocin – has immense control over how we feel and behave, yet it's routinely hijacked for commercial gain.

We already know this. We've been marketed sugar our entire lives. Yet it remains cognitively less taxing (and thus metabolically cheaper) to soothe affect – spoon ice cream into our mouths – than to dig into why we're perpetually dissatisfied. Another reason we're such easy marks is that we've been socialised to bypass our internal body–brain channels.

For example, we eat what a lifestyle app tells us to in pursuit of a body-image goal instead of sensing what our system really needs. It is possible to get an internal 'read' on nutritional requirements. But it's also tough to coax this dialogue back and clean up the signal once it's been scrambled by decades of dieting and calorie-counting (and resultant bouts of over-eating and self-loathing).

The same goes for the energy we expend. Movement's twentieth-century rebranding as exercise has burdened many with a messed-up relationship to not only body image and self-worth but also to why and how we move at all. The impulse to move is hardwired. In dancers, it's mandatory. The feeling of euphoria that comes once you invite the body to make every decision is one of sublime surrender – and the combination of exertion and emotional pleasure can bring hours of transcendence.

Even for those who do not move professionally, it's wild that movement, which every single body adores in early life, soon transmutes into a chore, bore, or a prop for self-flagellation. As nutritional, cognitive and sports science have proved, calorie counting doesn't tot up. No two humans – even identical twins – react the same way to the same meal or activity.

Still, given the nefarious tactics currently employed by social media companies, classic advertising seems quaint. As the surveillance economy's darker arts have been exposed, the extent to which they manipulate internal chemistry has come to light. That Facebook was designed to foster addiction via dopamine hits is old news. That it and many other social and media networks amp up arousal (whether in the form of love, want, hate or fear) to snatch attention and lock in engagement is common knowledge. Politicians have exploited affective responses for centuries, instilling fear, anxiety, and mistrust in us so that we vote and act defensively against a make-believe 'enemy' rather than demand entirely fair and free, let alone benevolent, societies. This is what authoritarianism is. What racism is. It's homophobia and transphobia. Bigotry. Bogus 'culture wars'. Terrorism.

Again, we know this. So, why is it effective? Why are intellectually competent, abidingly decent humans still falling for it? Because the body and mind are one – and this has precious little to do with intellect or reason. As much we might be aware that what the Home Secretary is saying is hideously xenophobic, a primordial fear response occurs when we are told 'you are not safe in your own homes.' Once aroused, that feeling is parlayed into emotion directed at whatever unfortunate target is being glibly exploited.

And until we develop a nuanced understanding of how our chemical states and emotions interplay, we remain open to being controlled in this way. If something you read online riles your body, before reacting and directing emotion out, pause, take a breath, and query whether it's genuine anger or a manipulated response: some corporate or political entity mucking around with your internal chemistry.

It's set to get worse, too. If it's irritating to be targeted by ads based on supposedly private messages, social networks, and browser histories – consider what might be on the horizon once Amazon *et al.* access your interoception. (If you let them.)

Technology in and of itself is, and always will be, morally neutral. Smartphones are not evil, nor are they good. However, as with relationships, if your attachment to tech becomes anxious, it's time for a break. Biometric techs such as Fitbit, Oura and Apple's Health app have been around for a while. It's cool to find out our stats; these devices can be lifesavers. But they're not wholly reliable. And for those with eating disorders or quantified-self fanatics (those with an intense focus on self-tracking to improving health), they can be psychologically harmful.

As for biometric-based technology such as Studio XO's XOX bands – it's fantastic that people who are physiologically or psychologically unable to match embodied feelings to emotions can wear a wristband and re-forge these links.

But it's also possible to one-click buy consumer-geared wearable devices that intrepret 'mood' based on biometric data. Without stringent privacy settings, the surveillance economy may well soon be plugged into your pulse, blood pressure and galvanic skin response, which it might correlate with decades of personal and social data to sell whatever they detect your body needs and/or what you'd be most likely to buy at that moment. And all potentially without you knowing how you feel.

Chilling. And a convincing case for paying close, off-line attention to interoception.

By sensing and regulating our mood and energy fluctuations through interoceptive awareness, we enhance our own physical intelligence. Practising awareness, moving mindfully, and extending physical skills foster continuous transformation of body and mind. Engaging all sensory pathways to perceive, interpret, adjust actions, and build versatility helps us avoid mishaps and acquire new movement patterns. Consciously controlling our attention, shifting from a broad to a narrow focus and between senses, improves our perception of surroundings, sensations, and emotions.

This journey of self-discovery and self-awareness isn't simple: it demands constant attention and significant retuning of our consciousness. But the rewards are immense. Once we've made this new awareness of our body and its states and moods a part of our daily life – the part of our inner self that it has always been, without our realising it – then we can begin to reach realms of heightened experiential, communicative and artistic potential.

PART TWO

Movement is Communication

7
We are Touch

How much touch is in your life? Do you prefer to touch or be touched? Do you ever think about the quality of your touch? Do you suffer from touch hunger? Touch loneliness?

Before humans developed verbal language as we know it today, touch – such as grooming – was principally how we communicated love, trust, and all the other pro-social emotions. It was how we bonded, mated, and parented.

Touch is our earliest sensual relationship. Indeed, it's still technically our primary sense – our 'first' sense – because we develop it in the womb at around eight weeks' gestation – initially in the face, lips and nose. At fourteen weeks, most of the head has developed touch sensitivity, and by twenty weeks the whole body has. When you're swirling around in utero, your foetal nervous system can react to touch from your mother. Touch helps us to 'see' before we can focus visually. Touch is our boundary sense. Internal and external. Given and received. Emotional and informative by design, it is our punchiest sense in terms of communicative prowess.

WE ARE MOVEMENT

When we are born, we rapidly adapt to bright light and loud noises and breathe air for the first time. Although this is an overwhelming experience, touch calms us. Our mother's touch communicates that we're safe, anchoring us in this strange new world. Stress reduction is the very first role touch plays in our lives. A relaxed, caring touch always reduces stress, no matter how old we are.

Once we are born, during those months when we are utterly reliant on our caregivers for our bodily needs, there is little, if nothing, that can be done without touch. We rely upon it as our bodies learn to regulate temperature, heart rate, and immune system through skin-on-skin contact with our mother's vital systems or our parents' embrace. Touch is critical to our healthy development and our ability to form stable emotional bonds. Only by touching can we learn to establish trust and care.

As we grow, our social touch history impacts our physicality and sense of self. Without touch, we suffer psychologically, and as body and mind are one, so do we physically. For those who live alone, the coronavirus lockdown delivered a chilling reminder of how much we (as a species) need touch to stay healthy.

In adulthood, we may encounter touch's phenomenal power to assist us in generating intimacy with another, communicating non-verbally (both well and otherwise), and releasing embodied and psychic feelings of tension, pain, anxiety, and stress; this is the bedrock of all body-based therapies, from sports massage to acupuncture, social dance interventions to shiatsu.

★ ★ ★

When Aristotle assembled his mind–body schema in his *De Anima* (*On the Soul*), he categorised skin and touch as lower senses within as lower senses within the entire sensorial hierarchy, regarding their role in contributing to knowledge and understanding, even as he also said that touch is everywhere and nowhere. He's right: it's nowhere when it's not activated or below conscious attention, but everywhere once it has been. Touch sits almost below the horizon of consciousness, perhaps more so than the other senses – its ubiquity, its everywhereness rendering it forgettable. It's often not until it's activated by pain, stimulation, or arousal that we even acknowledge it.

Our sense of touch constantly draws in information about the world. It never stops. It processes so much information from so many different sources that, in practice, the only way we can begin to make sense of it is to tune a massive proportion out again, which means that we don't think about the pressure our feet exert on the ground as we sit, or the force we're applying to a keyboard, or what it means when our elbow rests on a table. Even though we have several different types of sensory neurons crowded with specific nerve endings, each is tuned to their facet of the sense's capabilities.

Some aid proprioception by detecting skin stretching; others convey mechanical forces, allowing us to feel vibrations, edges, textures and so on; further sensors react to 'natural' stimuli, such as itches, chemicals, temperature shifts and inflammation. This data barrage sent from our touch senses remains primarily invisible to our conscious mind.

Unlike taste, smell and vision, which rest in specific organs, touch is embodied across many surfaces. Our skin, the largest sensing organ in our body (soft and pliable for maximum movement but also tough enough to resist tearing), operates like a vast feedback interface, receiving various messages updated continually. Signals from the delicate skin on our eyelids to the thick skin on the soles of our feet, from the nerve endings in our fingers to the hair follicles on the back of our neck, are mediated in a constant bio-loop of motion and evolution. This capturing of rich body data allows the skin to act as a cooling system managing our sweat, providing a wraparound waterproofing and the first line of defence against bacteria and disease. From the different colours, textures and thicknesses of our skin, the messages are plural and complex – the skin protects and shields us against heat, light and injury, as well as stimulates pleasure, chills and thrills. And we construct a holistic bodily idea of self because of this continuous polyphonic feedback mode.

Each of us needs to touch and be touched regularly so that we retain sensitivity to how it feels and the emotional responses that simultaneously arise from it. We have at least ten distinct touch-related body maps in our minds, and these are plastic.[1] These touch-related body maps are representations in our brain that help us understand different sensations. The somatosensory cortex processes touch, while the homunculus illustrates how various body parts relate to brain areas. Proprioceptive maps tell us where our body is positioned, and tactile maps show sensitivity to touch. Temperature maps detect heat and cold, while pain maps help locate discomfort. The vestibular map aids balance, and visceral sensation maps represent feelings from internal organs. Facial sensation maps focus on touch sensitivity in the face, and multisensory integration maps combine touch with other senses to enable us to completely understand our environment. Together, these maps help us perceive and interact with the world around us.

If these body maps in our minds aren't stimulated, they erode, which robs us of touch and, thus, communicative sophistication. In addition to losing dexterity when manipulating objects, we forfeit the pleasure touch can bring.

We also all have an incredibly sophisticated understanding of what a touch means, even if we are not consciously aware of it. Studies have shown that many emotions, those which are usually directed at others, such as anger and sympathy, can be communicated through touch alone. But this ability can atrophy if it's under-used. In one study – which illustrates what happens when you stop expressing a range of emotions via touch – men proved unable to detect when women expressed anger through touch. Meanwhile, women couldn't 'read' male touch that was intended to convey sympathy.[2]

By their profession, dancers touch and are always touched. Dancers spend years training in techniques based on touch, listening through touch to their own body and others. Physical touch is unavoidable in dance. From their student beginnings, where touch is (and should be) used as an accessible and safely efficient teaching tool, to all kinds of partner dancing – in classical ballet, tango, swing and contact improvisation – touch is an integral and vital part of the form.

The dance profession has come a long way in the past few years, prompted by the #MeToo Movement and the scandal of abuse in professional gymnastics, to reset its thinking around consensual touch in training and a professional context. This reassessment of consent with measures now in place to promote more equitable and respectful relationships in the studio after revaluating past power dynamics have been widely heralded. With more transparent and stricter guidelines for safer working environments now in dance, the potentially vulnerable daily negotiation around touch, choice, autonomy, speaking up and listening have been disrupted. And the dance ecosystem is better for it. Although still not perfect, these necessary changes have made a significant and influential modification to dance practice and culture.

With this in mind, we are discussing touch in this chapter with the understanding that consent is being constantly sought and granted through open and collaborative dialogue, with regular check-in moments as per best practice guidelines.

An Invitation to Touch

It shouldn't surprise us that the interconnectedness of touch and consent are prevalent in any conversation around people dancing together — although why it has taken us so long to figure out some safety regulations around this is another matter. Dance is, after all — arguably outside of sex — the most intimate act we human beings engage in with another person, sometimes with strangers, and therefore often stands outside the usual norms and conventions of interpersonal relationships. Take, for example, our habituated understanding of our personal space.

In psychology, the space around our bodies, regardless of where we are, is called 'peri-personal space'. Remember, we discussed the kinesphere in an early chapter – the zone of reach around your body, the way you extend your body into the world? Well, your peri-personal space is the malleable external border where we can be touched, as in reached by fellow humans and other entities.

Peri-personal space is shaped by the way our brain interprets signals regarding our body's state and its awareness of our environment. This process is guided by information from our senses – sight, sound, touch, and body position. Within peri-personal space, these senses are heightened, as the nearer someone or something is, the greater the potential danger should they or it prove a threat. And on occasion, they (it) will be a threat – an unwelcome invasion of your peri-personal space – assault, theft, projectile, bull, which sends your senses into overdrive and provokes your thinking body to react accordingly.

However, an invitation into our peri-personal space, accepted by our dancing partner, is quite a threshold crossed. Given our available options now for avoiding harm once another person is close enough to touch us, our perceived vulnerability is extremely high. The very act of partner dancing has permitted us, at this moment, to breach a conventional intimacy boundary. We welcome the disruption of our peri-personal space and settle into a new conversation of sensation through touch. There is no touch without venturing through another's peri-personal space. Not many activities between humans facilitate this crossover so easily. Dancing together, touch to touch, takes you into a whole new world of sensory experience. Physically thinking with another sentient being is not the same as physically thinking solo, and the benefits are enormous.

This understanding of peri-personal space and its relationship to generating intimacy through touch has been advantageous to me in working in countries that have experienced debilitating conflict or in environments where touch has been, in some form or another, weaponised. For many years, Studio Wayne McGregor has worked with the British Council in places where people have experienced extreme trauma – Bosnia and Herzegovina, Serbia, Romania, Belfast, Turkey, Siberia, Senegal and others – using dance and physical thinking as a tool for dialogue, exchange and often healing. Through workshops, performances and collaborative creative projects, we have travelled the world and been lucky enough to share profound and life-changing encounters with incredible people striving hard to mend their communities and themselves through positive interactions with their bodies and the bodies of others.

In 1996, we were the first major international performing arts company to tour Bosnia, Herzegovina, and Serbia after the war. We experienced first-hand the brutal toll, exhaustion and human scarring such conflict engenders. We were there ostensibly to perform, hopefully providing some inspiration and light in a way only soft diplomacy and arts that transcend borders can do. To gently remind the audience that they were still part of a broader world that wanted to connect and communicate with them. But for us, it was a significant opportunity to work physically with groups of interested non-dancers, have some fun with local community groups, and use the inherent power of dance to listen, engage, and play creatively.

It was clear at the beginning of each workshop that we were back to basics. There was too much tight tension, anxiety and distance in the room to start with speed, which is a tactic I often use to galvanise a group before they've time to judge or think. We had to go slow – re-acclimatise their tentative bodies to the wonders of touch and the relief of embodied cooperation. I reached for an exercise I often used when foraging for a way in. It is an exercise that asks you to enter the peri-personal space of another and, through a 'listening' guided touch, caress within.

Try this with your partners, your friends, your children, your parents, your boss, your work colleagues – strangers. It can be a revelation and much more challenging than it seems.

The exercise is a simple blind lead.

It begins with you introducing yourself to a partner.

However familiar you are with one another, please start by saying hello and making eye contact – the first point of touch.

One of you stands behind the other. The one in front shuts their eyes. Stillness.

Slowly, the person behind touches the small of their partner's back with their hand. It rests there. If you have your eyes closed, feel the hand on your back. Relax.

Gently, the person behind starts to walk, guiding their partner with their hand in the back.

Moving forward, use just enough pressure to keep them on the move. Pull back if their back disconnects from your hand. Keep the signal between your hand and their back 'live'.

Still, slowly, try changing directions, steering your partner right and left, releasing pressure when you need them to slow down and increasing pressure when they can safely go faster.

This is an exercise in building trust and confidence. Please proceed slowly and never deliberately steer them into obstacles. Take care of them.

If you have your eyes closed, please don't peek. Let your partner be your eyes. You only have to react to the signal in their hand.

Start to feel the difference in pressure, direction, speed and force exerted by the hand that guides you.

Are you breathing more calmly now? Are you less tentative in the legs, no more pushing back against the flow of the hand but with it, in tune?

Be braver now — both of you. Change speeds, change the direction of travel, vary the signal, and thereby vary the flow. Stay connected. The hand touching your back is the guide to the vastness of motion. You are touch-ready.

And come to a pause. Swap over roles. Remember to start slow.

The dexterity of a hand, the minuscule movements we can make at any given time with this beautiful instrument of complex coordination, is constantly incredible to me. Here, the fragile connection of hand to lower back provides a conduit to motion, to shared coordination, to our first dance together.

Feeling the weight of that person in your hand, sensing them, and giving you permission to take them anywhere in the room is liberating. Realising that you are responsible for their every movement and safety as you tune your ability to drive them nonverbally is also liberating. Equally, giving in, receiving the touch signal in the back and going with it, trusting in another to keep you in flow, and believing that your journey will be without incident is profound. This 'knowing' is the first step to a genuinely listening body.

The adventurous individuals on that day in Bosnia, those special people in that room who had existed in a state of hypervigilance for years – no relief, no reset, and often in this state as a matter of life or death – were alert in the most burdensome way. It took them time to recover any sense of meditative trust or listening to their bodies. But their catharsis was so visible that we all stood and cried when they did.

Our tears were tears of joy. Through experiencing the touch of another, with this human connection made and divined between two listening bodies, we not only moved together, but we were moved profoundly.

Have We Lost Touch?

In the dance form known as Contact Improvisation, dancers practise an advanced version of this blind lead. The skill is to improvise while staying in touch with one's partner's body surface, to 'listen' to what their form is saying and respond, rather than impose direction or intention, all the while flowing with gravity and the inertia of bodies themselves. This live haptic negotiation is when bodies communicate through manipulating weight and force, pressure and release; they choreograph their expressive journey together by exploring the touch qualities of hardness and softness, lightness and heaviness.

Contact Improvisation is a joyful, thrilling and pleasurable dance genre, both as a participant and as an audience member. Much like its more straightforward relation, the blind lead, it can reveal how we can use the touch sense to change our relationship with and understanding of the world around us. Outside of the professional dance world, Contact Improvisation has also proven a worthy therapeutic tool to re-accustom the touch averse, whether suffering from PTSD or social anxiety. It reveals how we can use touch effectively to generate newly discovered dimensions of intimacy between us, thereby harvesting the positive sensations that safe touch elicits.

But so many of us are out of practice with touching and, in the process, have lost touch with our bodies and ourselves. Culturally, this relates to the atomisation and migration of our tactile and social experiences to online life, almost completely ignoring touch as a sense that we should continue to cultivate.

During the pandemic, it was striking how painful it was when our habitual experiences of touching and physical contact were curtailed. That primary sense we had until then was so easily taken for granted. It wasn't only that people had stopped hugging each other and interacting socially; many were robbed of the ability to use touch as a communication channel with their loved ones in heightened situations where no words would ever be useful. The kids who were able to see but not able to touch, console and comfort their dying parent; those heartbreaking scenes of hands touching either side of panes of glass, or whose last encounters with loved ones were mediated through Zoom. We were reminded of the power of touch and celebrated when we could touch again and feel one another on a visceral level. But how long did that attention to touch last?

The amount of physical contact most people experience daily can be meagre. As we age, the touch quota generally decreases. How would you fare if you were to audit your touch income and expenditure currently? Yes, if you were to count all the objects, tools, surfaces, pieces of clothing, and architectural features like handrails and stairs your body has touched today, the tally would be enormous, but people? How many people have you had a touch experience with today, deliberately or accidentally? And for how long did the touch last? Please take a moment to jot it down.

You will likely be in a touch deficit. Either you hardly touched anyone, or hardly anyone touched you either in an intimate way or indeed a non-intimate one. And if they did, it was only a brief moment of contact – a handshake, pat on the back or, at best, a friendly hug. Why does this audit matter?

Back to our skin. Skin sensitivity depends on the stimulation we receive. The less stimulation, the less responsive our skin becomes to touch: literally. Because our skin cells completely regenerate in about twenty-eight days with individual cells potentially replaced in as little as two to three weeks (the average adult has some 4 million touch receptions in the skin), the less the skin is touched, the fewer receptors are replaced and the less sensitive to touch we are. Our bodies don't waste energy on creating cells that are not being used. This is a case of use it or lose it.[3]

The opposite is also true. The more we expose our bodies to stimulation, touch, and pleasure, the more pathways are created in the brain and nervous system for these sensations. The distance of experience between that blind lead exercise and the full-flowing, holistically realised technique of Contact Improvisation may feel overwhelmingly enormous – it will take hours of practice and group interaction to attain anywhere near the skill level of professional dancers' effortless touch fluency. But even small steps towards improving your touch pathways and your touch geography will enhance your tactile communication and receptivity.

Start Simply

Re-establish a new and beneficial relationship with the ground upon which you walk. I fly a lot – hundreds of hours on aeroplanes crisscrossing the globe. Flying is a complex experience for me: it offers both a protracted period to have some time to myself and the dislocating ordeal of being untethered from the world, stuck in a claustrophobic tin tube with no escape! As soon as I get to my destination, to the hotel, quicker if I can – I take my shoes and socks off and feel the ground. Land.

That feeling of connectedness, of being grounded on soil or grass – touching nature – is awesome. It's qualitatively different from 'landing' on concrete surfaces, tile, or carpet. But after flying, any floor feels better than none. The best is when I arrive at my house in Lamu and feel the connection to the sand and the sea – simultaneously – walking along the shoreline and feeling the Earth's surface ebb and flow beneath my feet. Barefooted bliss.

Many people rarely touch the ground with bare feet – perhaps a trip to the seaside, or some occasional yoga, might be their only yearly experience of foot liberation; otherwise, feet are almost permanently encased, closed in, even at home.

But children spend years barefoot and free before we buckle them up. Experts encourage it: for a child learning to walk, shoes may interfere with how they use the muscles and bones in their feet and alter their gait. When walking barefoot, children receive feedback from the ground, building their awareness of their body in space. And that changes as we get older, precisely because we don't allow our feet to be free.

I spend much time barefooted at home, and outside when I can be. When I rehearse in the studio, especially if I need an energetic boost or a quiet centring, I ground myself by taking off my trainers and staying foot-naked for the rest of the day. Like many of the physically intelligent skills we highlight in this book, we lose the benefits of barefoot walking if we do not do it regularly. These benefits include better control over the foot's positioning as it hits the ground, balance, body orientation, alignment, and relief from improperly fitting tight shoes, which can cause bunions, fungal infections, and other foot-related issues.

You can begin with short 15-minute sessions to help your feet adjust gradually. Walking barefoot on grass, sand, or soil may reduce inflammation, lower cortisol levels, and alleviate stress. Some believe that the Earth's negative charge transfers electrons to your body, helping to neutralise harmful free radicals. And that's all part of the bigger picture about touch sense. If you take responsibility for your skin – that unbelievable sensing organ – you will water, exfoliate, moisturise, and treasure it. And that means taking responsibility for *all* of it – do you wash your legs in the shower? Do you ever pumice stone your feet? Or do you usually only touch your face, hands, torso, and arms? Neglect your skin at your peril – the skin is your conduit to sensation, your invaluable sensor of touch.

Touch Intimacy

As we enhance our tactile receptivity, we can change our relationship with touch and gain a more sophisticated understanding of how to 'use' it. How does one touch better or more confidently? Can we improve our ability to communicate through touch and get more of it in our lives? This choreographic exercise might offer a way in.

Apart from making up movements, teaching dance phrases and composing, choreographers spend countless hours watching people and how they move and interact with the world. Over the years, I have developed a way of watching that straightforwardly breaks down action into digestible attributes: the *'what, where, when, how, why'* system that we've already been introduced to. This helps me create new dance forms and innovate unique physical languages, but it can also help you open a whole invigorating world of experience through touch.

In the blind lead exercise earlier in this chapter, we invited the group to touch with *what*? And *where*? To touch with their hands on their partners' backs. The *what* is the hand, and the *where* is on the back. Through practising this one form of touch interaction, the participants explored what examples of non-verbal cueing and non-verbal communication were possible when the hand and the back meet, opening a channel through which novel sensory pathways emerge. By changing the *what* and the *where* in an encounter, we have the chance to practise a new sensory path and, therefore, enrich our internal touch map.

What if instead of using the hand, we changed the *what* and offered the wrist, forearm or back instead? We have many body parts to choose from; you can be as adventurous as you like (shoulder blade, thigh, foot, neck, mouth, cheek).

Use your back to touch *where* – the back of your partner. Can you still guide, start, and stop the motion and be your partner's eyes while using your back as the steering mechanism — just like in the blind lead with the hand? What feels different? What other bodily listening skills must you deploy to maintain that connection and sensitivity? Tick — a new pathway. Try another.

Changing *what* body part you are prioritising in any given interaction and *where* you are touching the other person gives you an ever-evolving menu of touch possibilities. Each new relationship adds a map to your personal touch bank.

In Contact Improvisation, the *what* and the *where* possibilities are endless, not least because dancers are as much upside down as they are vertical. But importantly, a 'no hands' dance is actively encouraged. Practising without hands frees the improvisers of the 'grasping/grabby' and controlling nature of our habitual hand interactions as they learn and become more aware of how they can use different parts of their whole body to touch. Hands are not banned per se but used in a symphony with other body parts. Letting go of rote touching models and re-evaluating the touch hierarchy can be liberating.

It may be worth applying this *what* and *where* technique in your most intimate encounters, too. It will activate your and your partners' sensory systems in surprising and delightful ways.

The timing and speed of your touch — the *when* — are also critical aspects of its effectiveness and potency. In both the blind lead exercise and advanced Contact

Improvisation, *when* is vital to the quality and nature of that interaction. There is coordinated turn-taking and smooth interaction management. These skills are activated when you genuinely listen to the other person's body and respond accordingly.

Much like in conversation, some people already have their next interjection prepared before the person speaking is even halfway through their sentence. They impatiently overlap with their following comment, diverting the dialogue to mono. Their insistence switches the faithful listener off. Try not to be that person when it comes to touch. Touch is a movement thought, and that thought needs time to be expressed fully. Practice following and leading interchangeably, refining your non-verbal ability to cue and be cued, to stay in a movement thought for longer than you expected, and to be ready to slip between these modes easily, with no speed bumps. Attending to your partner this way, focusing on timing and generating continuity, motion and momentum, is an essential cornerstone in your touch repertory.

Qualities of touch – the *how* you touch (and are touched) – are subtle competencies likely most in need of rehearsal, or at least expansion. Learning to vary the essence of your touch and nurturing your dynamic range is a considerable part of dancer training and skill-building. Recognising how we can mediate between tactile dimensions and actively surf between essential qualities of touch is paramount to more expressive communication through the body.

As an experiment in exploring the qualities of touch, let's start with our own body (and progress to others' when you want) – self-touch is a brilliant way to practise. And we will start with the fingers, aware that we can use any body part in any combination for this sensory journey. Take a finger for a walk.

Palms up, I take an index finger and touch my other hand as softly as possible. I trace each finger in turn, around the wrist, behind the hand, and trace its back before navigating my way to the inside of my digits. Up and down, I follow the curves of my fingers, the in-betweens – a miniature roller coaster. I repeat the same pathways but harder now, with more pressure. I can feel the bone and the muscle move, and the journey of my finger becomes less fluid. I stop momentarily and pause, my finger resting wherever it can – I press the spot, digging in and releasing. I go back to a gentler touch, back to where I started. That was hard and soft, slow. If I close my eyes now and imagine the same journey of the fingers, the gentle beginning, the same route, the stop – there is a faint memory of sensation on my skin. I can feel the touch without touching.

Keeping our hand map in mind and touching again, I repeat the journey of the finger only in straight lines, tracing NYC-like street grids along its surface. I still move slowly but with right angles, linear pathways and etched directions. The lines traced on the back of my hand feel different from the lines on my palm – thinner skin, perhaps. Back to the beginning, I begin again, but now all curves and circles. Take your time. Can I keep a continuous motion while redrafting my journey with a finger in hand? Try smooth, pendulum flow – now swaying to and fro? And home. That was linear and circular, sustained.

Try sharp, try fast, try long, try short.

Try hot, try cold.

Try… ?

These qualities of touch are infinitely combinatory; to hone them, you only need time in situ and practical experience. Similarly, by re-combining the *what, where, when* and *how* of touch (we have looked at them above in isolation for clarity), we can make these elements sing in harmony. As we experiment with where we touch together with when we touch, what we touch with how, this panoply of options begins to emerge and our sensory maps widen.

This widening and exposure to touch experience is an important foundation for enrichment, not least because touch releases valuable oxytocin, the feel-good hormone. All touch is conditioned by intention – the *why* we touch – and this directly influences the nature of our touch action – the *what, where, when,* and *how*.

Are we touching to console, humour, control, soothe, save, annoy, share, or pleasure? What kind of touch does this elicit? Our intentions – our purpose and meaning – are revealed (and yes, sometimes deliberately masked) in that moment of contact. When touched, we register the nature of that touch, its pressure, speed and direction, and interpret its meaning. We judge its emotional temperature and translate its literal weight, and we react (or not) with an understanding of its inherent content. In this way, touch is always an intimate exchange of meaning.

Intimacy, and how it's fostered and nurtured in a relationship (whether romantic, social or familial), is a profound and precious virtue of humanity, and nonverbal communication — eye contact, gesture, postural openness, facial expression and touch — are its primary vehicle, the *why* of touch. Being touched emotionally, positively, by any means binds us closer. Affectionate cues sent and received by bodies sustain and nourish intimate thoughts and feelings, encouraging intimate words, which spark affectionate cues — layer upon layer upon layer.

Humanity's sensitivity to touch — to its sensuality and intimacy — renders it a potent instrument of connection and communication. In a renowned psychological study, the Rubber Hand Illusion, participants watching a rubber hand being stroked, while having their own stroked at the same time but outside of their field of vision, soon begin to feel that distant touch on their skin even when they were no longer being touched themselves. They're tricked into subsuming the fake hand into their body schema. Similarly, viewing touch can ignite body transfer, so it's favoured to initiate 'presence' in virtual reality environments.

This is a powerful example of 'vicarious touch'. This ability to viscerally and emotionally react to touch upon others as if we were being touched is a remarkable projected attribute of our physical intelligence. For the socially isolated, watching skin-on-skin contact on screen is may be recommended as a balm for touch hunger.[4] When watching dancers (look out for kinaesthetic empathy in another chapter), actors or adult film stars touch, our personal touch narrative, inextricable from our most intimate encounters, at times seems to dance beneath our skin, enlivening us as we recall sensual and emotional experiences. As the curtain rises on any live performance, all this connective potential awaits in every audience member. This is the power of touch.

8

We are Presence

Carlos Acosta has a real presence on stage, in the studio, and in life. He has charisma, confidence, and effortlessness about him. He is 'comfortable in his own skin'. Wherever he is, no matter the company, Carlos holds the room. It is as if his body extends beyond itself, reaching out to occupy a greater volume than his limbs ought realistically to allow. This aura is attractive and compelling. It is a vital attribute in the composition of his physical signature.

Is this presence a natural state or a learned one? Can you improve your personal presence? Why would you want to? Being present is not the same as exuding presence, although together, they are intoxicating. Carlos definitely exudes presence, but it is his *being present* that impresses.

Being present is the ability to respond to the moment, to have one's senses attuned 360 degrees, to be able to improvise in real-time, and to be committed to the enjoyment of now. To be in-body, 100 per cent here.

How do you know you're present? That feeling might arrive when you dance, ride a horse, or hit your best stroke in a tennis game. In the physical and mental connection with others, we can experience the most heightened sensation of being present. Think about sex, your most intimate interaction with another person, where your whole body is alive to their touch and where your touch enlivens their entire body. Where our physical signals cue and guide embodied responses and where we are listening with and through our bodies, reaching a state of not thinking. The voices in your head are muted, and you find a complete and mutual flow state: that whole incredible sensorial buzz like a dervish moving to a transcendent state.

States of Presence

Patsy Rodenburg is a phenomenon — a voice coach and authority on states of presence. Her seminal teachings on mastering the art of being present are universally applicable. She often teaches using Shakespeare as the vehicle for exploration, and her work has engaged her with celebrated actors, musicians, correctional officers and political leaders. Her ideas have been transformative in my dance practice, and I have invited Patsy to work with me many times and in various contexts.

As Patsy puts it, 'everyone has known the feeling of being present. Babies and toddlers live there almost constantly. Great performers work in this state. Great athletes win in it. Great teachers teach in it. Every great communicator speaks from this place. When fully present, we do our best work and make our deepest impression on others.'

Patsy believes we operate within three circles of presence: Circles One, Two and Three. We need to utilise all three and can learn to move rapidly and effectively between them. Her theory hinges upon two essential physically intelligent practices: a solid command of attention and the ability to detect and transition between varying states of energetic arousal and embodied mood.

Circle One
In Circle One, energy does not travel beyond our introspective state. It is an energy that falls back into you. Shoulders sag, voices barely rise above a flat murmur. Still, this Circle has its merits. We all return here from time to time – to gather thoughts, process experiences, interrogate creative ideas; to improvise, eyes closed, hearing the body alone; to tend emotional wounds, quietly restore. Deep in Circle One, we cannot attend beyond ourselves; rewarding interactions are inaccessible.

Unfortunately, while clinically depressed, grieving, or otherwise existentially stricken, this state can feel interminable, an irreversible theft of personality. And while feigning a Circle One state can serve well as armour against unwanted attention, when we exist there in the long term we're an energetic vacuum – whomever we encounter will depart at least a little depleted.

Circle Three
Circle Three, by contrast, is forced projection, torrential bluster. Third-circle energy is energy that is pushed out. It's energy that looks to the future. It can be embodied by people who barge into your space with their chests thrust into the air. Their voices boom, and they snatch all the oxygen out of the room. It's controlling. Think dancers who fix their gaze on an imaginary friend at the very

back of the auditorium, blasting energy at them in fits of Annie-on-amphetamines positivity. It can be a woefully inauthentic state, as demonstrated by peak-Three: Trump. It is amplified pomp and huff – an insecurity-fuelled steamroller. And an attentional black hole: the expectation is all eyes on me in Three!

Circle Two
Second-circle energy involves an exchange. That could be between two people. Or between a person and a toy, a piece of music, or a book. You are connected to it, and it is connected to you. There is give and take. You are present. The energy is equal. We are all born in the second Circle. Children are alert, partly because they must be. Second-circle energy is about survival. It's also about intimacy. You cannot be intimate with someone else unless you are both present – absolutely there, in the moment – with each other.

An experiment: notice how people communicate presence/being present in their Instagram feeds. What's their presence default, their bias? Are they communicating in state one, two or three?

Life pushes us out of the present. It happens slowly to children. We lose something of our ability to exist in the second Circle. And yet, we all can bring ourselves back.

Finding our Communicative Centre

Thankfully, in and of itself, the essential act of noticing which Circle we are in at any given time starts to tip us back towards our communicative centre, for it is only in Circle Two that true presence — innate charisma — is revealed. Here, we feel at ease, embodied, connected to our environment, and alive to companions. We flow freely in a vocal, physical and musical dialogue with ourselves and others. This doesn't make socialising hitch-free or instantaneously dissolve relational tensions. But it does grant us a tried-and-true initiation point from which to navigate ever-shifting social climes. A trusted place from which to advance, knowing that, come what may, we're adequately prepared to negotiate it.

The three fundamental states of presence influence how we communicate and receive our messages. They condition our choices and affect our sense of self (of being present or not). Because they are identifiable (we can learn to recognise our biases), we can 'play' them at will. Slipping between the states, practising being in them, and moving elegantly out of them is one of the most revelatory and powerful instruments in the physical intelligence canon.

I recently witnessed an everyday example of this 'play' while moderating a Q&A session with Carlos in Venice. Carlos is a physically relaxed kind of guy. He is very easy with himself and in his surroundings. Even in a public speaking setting, in interviews, he sits back, casual (a cool Circle 1). A young Mexican dancer was waiting to ask his question, and when it was his turn to speak, he tentatively shared an emotional, difficult, and personal plea for help and advice (a very different, tense kind of Circle 1). Carlos instinctively realised that it wasn't appropriate to receive such a heart-felt and probing question in his state of almost loose relaxation. He gently sat up and forward, grounded himself, and in doing so, he fully opened up his body toward the questioner, his eyes soft and attentive — an apparent move to Circle 2 and an invitational cue for more.

Without saying a word, purely by offering a state of presence change into Circle 2, the young dancer read Carlos' non-verbal affirmation and moved (both physically and psychologically) into a more unrestrained mode of communication. Now, in Circle 2, the questioner released himself: an accumulated mix of doubt and conviction filled the air. Finally, this expressive torrent found a space to release until our dancer inhabited Circle 2 confidently and with gravity. The conversation flowed in a tsunami of ideas and exchange. Carlos was clearly moved; I could see a tear in his eye.

Throughout, I doubt that Carlos was ever consciously thinking 'what am I going to say?' or 'what should I do next?' or 'how should I move?' — his was a felt response. His expert physical intelligence, refined through his years in dance, naturally matched his emotional intelligence at the moment. Carlos knows that every movement he makes, every state shift he executes, is impactful. He brings into his awareness a read of the current relational context to positively affect change. We cannot all have his experience and skill immediately, but we can work towards more fluency through practice. Merely noticing is an excellent beginning. Ask yourself, how am I listening right now? What signals am I receiving? Is there a way I could listen differently through my body that will facilitate a more vivid conversation?

Sometimes, I find myself in encounters and conversations where I feel resistant; my whole physicality builds a wall. It may be a direct result of the content of the conversation and the words themselves, but it may equally be influenced by my intuitive reading of the person. I harden, collapsing my entire presence inward to a Circle One state. I can choose to remain there, uncommunicative, unlistening, and blocked, or I can shift my presence to Circle Two, hoping that we can undertake a re-read or perhaps start again and soften into correspondence. Correspondence is always worthwhile. A genuine state of being present is always a state of facilitation, a kind of channelling that unlocks the path to a deeper and more meaningful connection with our fellow travellers. We can learn and grow by observing the best people who comfortably inhabit the present. Observe, for example, how Barack Obama's physical intelligence came alive when he met elderly civil rights activists, making them feel immediately at ease. What does he do? How does he move, and what states of presence is he navigating? When does he shift, and why?

Channelling Presence

Outstanding movie and stage actors master this 'play' of presence to fully embody the characters and the roles they inhabit on stage and on film. They must channel skills in recognising, enacting, and 'feeling' by proxy the presence states of others to express the complexities of emotion through their bodies – to convince audiences, making them believe that the 'artificial' moment they are simulating is actually happening, and that they are not, in fact, actors at all.

How do actors jump into their characters' bodies, minds, and motivations? And how do they truthfully inhabit these diverse roles?

As a movement director, I invited Patsy to my rehearsal and training with the actor Alexander Skarsgård. We were working on developing Alex's physical presence for his role as Tarzan in the Warner Bros blockbuster *The Legend of Tarzan*. Tarzan, a man brought up by gorillas in the jungle, has 360-degree senses and animalistic alertness. Alex had already re-sculpted his body with admirable discipline and dedication and was working hard to embody primate behaviour. We had spent hours analysing monkey videos online to note their motion and understand their wider instinctive and communal habits. We had advanced to practising primate reaction, locomotion, and simulating monkey behaviours. Alex was adapting well and intuitively responsive to any physical challenge we placed in his way – which included a primitive but thrilling primate gym.

Next, we sought assistance from Patsy to connect this visceral skill set with the text that Alex had to perform. The physical and vocal elements needed to merge sympathetically on screen to appear believable. Patsy shared her theory of three states of presence, where voice and physicality are regarded as one (vocally, Circle 1 is whispering, Circle 2 is conversing, and Circle 3 is shouting), and we practised moving and speaking fluidly between these states.

★ ★ ★

How could his vocalisations, sounds, and grunts carry energy – how could they be made as kinetically resonant as the body flying through the treetops? Where would the sound come from in his body? How did it vibrate and modulate in varying scenes? How did Tarzan's voice change when forced to wear Western clothes? How did the body adapt to produce the sound? It became clear immediately that Alex could direct and nuance every aspect of his performance by attending to relative states of presence and gliding between them. He could play the scenes in myriad ways, shifting his attention along the presence continuum at will; he could experiment within the framework and invent authentic yet surprising acting choices. He could be spontaneous in an option space filled with new potential. He could be fully present as Tarzan both physically and vocally.

As I watched Alex fluidly slalom between the three circles, we observed a noticeable change in his physicality. His body morphed and adapted as he shifted his attention through the various circles of presence.

The impact was particularly strong when Tarzan's vocal Circle of Presence aligned with his physical Circle of Presence. For instance, when Tarzan stood confidently on the cliff edge in physical Circle Three and howled his animal call in vocal Circle Three, this synchronisation creates a more powerful and realistic embodied image.

However, Alex sometimes mismatched his vocal expressions with his physical actions during play. For example, when he gently offered his ape-shaped hand to his mother ape – physical Circle One – he grunted in acquiescence using his vocal Circle Three. This was an interesting choice as the disconnection between the vocal and the physical states of presence introduced an extraordinary ambiguity of meaning.

Was this deliberate? Not always, but it highlights the incredible potential for eliciting nuance and subtlety in communication when you bring awareness to your states of presence and effectively mediate between them. Sometimes, aligning your vocal and physical circles of presence may be useful to convey meaning clearly. In other situations, a more flexible interplay between the three circles (and the vocal vs. physical) may yield or reveal alternative truths. This is a form of physical and vocal subtext that naturally permeates all our interactions.

Sometimes in life, our presence cues are crystal clear and easily read, while at other times they are conveyed more subtly and ambiguously. Are we aware of this difference? Can we utilise the 'presence' recognition skills to effect and support what we want to communicate? Can we discern a mismatch of presence in others that reveals uncertainty or doubt in their statements? Alternatively, do they project a fulsome confidence that instils trust and confidence in us? We can prepare for any interaction or communication task by enhancing our ability to mediate presence and convert ambiguity into clarity.

Invite The Audience In

Sharing these ideas about presence with ballet dancers has often freed them from a particular tyranny that dancing in large-scale lyric opera houses can unleash. Perhaps it is ingrained in the balletic training that artists must PROJECT, amplify, push their energy to the back of the auditorium – a forced and forceful grace. In many ways, ballet as a form is designed that way: set flat behind the proscenium, the physical language is turned out, linear, and presentational. We could describe it as Circle Three dancing. But the most extraordinary ballet dancers, for me, utilise and massage this language as an expressive and communicative tool and invite you closer; they meet you energetically halfway between the stage and the auditorium to commune in a safe space of the we. For the audience, this seduces you into the world on stage as an intimate collaborator in the dance rather than pushing you back, forcing you to watch from an objective distance.

This is, of course, a meeting in Circle Two, the place of exchange and mutuality, of with and for. When you witness a dancer slide from Circle Three to Two or indeed One to Two, you witness a quality of action that compels you; you immediately know that the performer wants to let you in; they are seeking a connection, your undivided attention.

On stage, when you're in Circle Two, you can feel an audience, feel individual members of the audience. You feel what's happening to them as you dance; you feel where they're sitting; you feel whether they're on the edge of their seats, expansive, tight or relaxed; you can sometimes feel when they are breathing as one with you. This audience presence, this out-there organism, can also directly affect how the show goes. The coughing, the fidgeting, the clapping, the whispering, the cheering, the silence, the (im)patience of the whole sends shockwaves backstage that we ingest and receive. Our instruments on stage, our listening bodies, are as attuned to the audience's attention as we are to our company's signals. Audiences have signal-sending bodies, which emit and receive. And because of this live dialogue, this conversation across the floodlights, one can effect change in both directions during a show. As susceptible as performers are to audiences' presence, performers calibrate their performances to affect the audiences. The performance may need an energetic lift, more audience eye contact, or a concentration on intention. Returning the dance to a state of the now, less rote and more alive, will re-engage the fidgeting few. Sensitive dancers notice this and exercise their agency to affect the change – they can strategise, convert, and influence. We can kinaesthetically nudge the audience back into now from a disconnected apathy (a scary Circle One) or a boisterous impatience (a noisy Circle Three). We can restore that lost connection.

I am increasingly fascinated by these bio-loops between audience and performer. Current experiments, testing retinal tracking, breathing monitors and biometric sensors on audiences, offer simultaneous live feedback to performers that may be incorporated into a show as it progresses. Imagine capturing a passive audience, recording low states of adrenaline: what might the performers do to raise the level of engagement? It is even possible now to re-compose dances live in order to recalibrate audiences' perceptions as they are collected, altering the fabric of the show itself. This kind of interactivity, presence mediated by technology, is the new frontier for theatrical presentation.

WE ARE MOVEMENT

Unleash Your Creative Power in Circle Two

The quality of being present and inhabiting Circle Two, that space of reciprocity and immediacy, is for me most important when I'm creating in the studio. Except for my quiet morning ritual, almost all the activities that make up my days are ones where I have to stay in Circle Two – working collaboratively in a dialogue with others. It can be demanding. On those occasions when I have a six-hour choreographic day, and I have groups of dancers coming in one after another every hour, I cannot afford to be thinking about what I was doing yesterday, or an hour ago, or what I'm going to be doing tomorrow. I must receive those artists with all that they bring in that moment and try to stimulate an inspiring creative act between us. If I'm too heady (Circle One) or too forward-thinking, driving to get to the end of something (Circle Three) – if I'm too keen to finish the phrase, the section, the composition, the work – I don't get richness in that Circle Two encounter.

There are practical things I can do to stay in the zone and techniques to reset myself and the incoming dancers, but it usually starts with priming the room. Mostly, I do this with music – I select a track. Never the score I am choreographing to initially, but something else. I need music that changes the energy in the studio, shakes me out of the last choreographic encounter and into a new one and suggests something to the dancers entering the space of the 'feel' and expectation of our session to come.

Choreography is 80 per cent the psychology of preparing dancers to work with you collaboratively. To settle into an alive, bi-directional state of presence is as vital as the moves we share. I'm acutely aware that when dancers enter a studio, they come in alone – no cellos, paintbrushes, or scripts. They enter as themselves, repositories of their day so far, in all their Technicolor vulnerability, neurosis, imagination, energy, and complexity. I am the same, full of the contradictions of being human. My energy of the day meets theirs. I'm not a static entity in the studio. Perhaps that day, I had bursts of creative insight in the first hour, engaged in nuts-and-bolts grafting in the second, vamped for the third and now, in my fourth hour of making, faced with a team of fresh dancers, all of us wonder what our time together holds in store. We negotiate immediately, finding a way of entering this creative act together. Part of this negotiation is setting aside pre-conceptions about yesterday and the work you have already made together and beginning with a clean slate – the latest reading of that person as they are right now. People are recomposed daily, yet we too easily confine them to type. We are too quick to categorise – this is the funny, quiet, or tricky one – and use this static render as the basis for all of our interactions.

Let's take one particular dancer, Joe. I never expect Joe to come into the rehearsal the same every day – in the same mood, energy, physical condition, and presence. Part of my job is to work out what his constellation of atoms tells me then and there and use that information to fuel a productive session. To ease Joe into his best physical and imaginative state. Then, we use our hour well: we play, discover, and invent. So, as Joe walks through the door, I'll do an agile scan – a 'read' of him, searching for the instantaneous clues to his now. What is he communicating physically? What is his posture telling me? How tired is he in the face? Is he smiling? What sort of eye contact is he making? And I will always check in with him – a moment of warmth or humour from me, a question as an uncomplicated welcome. These precious moments are some of my favourites in the rehearsal process. They provide a connection that enlivens me as much as they help to balance the forces in the room. They offer a time to let go of the past (even the recent past) and dive into the present tense. All that matters for the next hour is thinking physically through and with our bodies.

Once someone is within visual range, we generally form a perception (and, inevitably, a preconception) from their body's position in and movement through space. This primes us regarding their physical condition, intention, mood, and character. The daily scans I perform in the studio are simply elevated versions of the 'reads' we do on people constantly. The better the read, the better the potential outcome. Critically, the skills to respond persuasively to the information the read offers us, perhaps facilitating a change of presence, are tools we can adopt to help improve all of our communication.

As well as being revolutionary in terms of self-awareness, mastery of those circles of presence, especially the learned capacity to be 100 per cent present – to coax, breathe, or dance yourself into Circle Two before greeting others – will entrain a versatile social lens through which to detect where the human you're meeting is actually meeting you from.

This presence play is something I use continuously to free the body (and voice) to be more authentic. I owe a huge debt of gratitude to Patsy for her wisdom.

9
We are Gesture

My background in dance and choreography has taught me much about gesture and other nonverbal communication. Dancers 'speak' with their bodies (as we all do) and often use refined gestural language, part of a broader arsenal of techniques, to tell specific stories or convey meaning from the stage.

Dancers become attuned to what every gesture they make signifies and how each of their gestures can interact with another gesture on stage. They practise, develop, and evolve skills that expand the palette of possibilities when they dance as characters: it's one way in which they transmit complex emotions through the body to the audience and embody a role.

There are many expert storytellers in dance, from the unlikely cool of Fred Astaire to the passionate brilliance of ballerina icon Lynn Seymour; these dancers have understood that their varying uses of gesture are a large part of an audience's comprehension of their character's traits, status, motives and state of mind – gesture making visible the interior complexity of their character and making them believable, truthful. To believe is to go on an extraordinary adventure in the theatre, a unique version of immersive reality where you, the witness, are submerged into worlds of the past, present and future and feel as the characters feel – right there, right then.

Bodies tell different stories in every encounter and communicate an enormous amount about our personalities and histories. Our past experiences affect our gestural language. Pain, joy, and fear all build up like sediment and are the foundation of our relationships. In the same way, our bodies hold the histories of our past social interactions and relationships. This is a record, an audit trail of our connections and relationships with others, one that can help us understand and forecast our performance in similar situations.

This chapter aims to unpack the processes that allow us to convey thought and emotion effectively through gestures. We will interrogate the ways in which our gestural language sometimes controls how we are perceived by others, irrespective of what we intend to express. Hopefully, it will encourage you to expand your own repertoire of gestures.

Reading Hands

Besides our faces, our hands are the most evocative and versatile expressive tools we possess; they contain the highest concentration of bones, twenty-seven in each hand, and reach the greatest speed we can muster. Who would have predicted that a finger snap occurs in only seven milliseconds, more than twenty times faster than a blink of an eye?

We now know that multiple brain areas are tuned into the perception of hands and their motions. We translate hand movements and the dynamic of hand and arm gestures, as well as their perceived aggression or softness. When we combine all this information, it helps us understand whether the person communicating with us intends malice or kindness – a salient precursor to any interaction!

In a foundational 1969 study of human nonverbal communication, Paul Ekman and Wallace V. Friesen categorised gestures into five classes of motion, three of which are relevant to the hands: adaptors, illustrators, and emblems. Fifty years on, that trio of categories remains extremely useful for reading people and for understanding how you communicate gesturally.

Adaptors are touching gestures that indicate internal states, such as anxiety, an outward manifestation of stress, or boredom. They're a prevalent gestural style, and you probably (perhaps without realising) display a range of adaptors daily. Examples include fidgeting and foot tapping, clicking a pen, and twirling your hair.

Adaptors can be easily classified in three ways: self-adaptors are the self-comforting gestures we make, for the most part in private. Grooming activities such as tugging on your beard, biting your nails or scratching that itch for an excessively long time fall into this category; object adaptors involve something other than your own body such as paper clips, pens, drinking straws, and cell phones, which can indicate levels of boredom and anxiety. It's easy to pick up on these tics, and alter adaptors are made in response to another person, folding your arms in response to a telling off, crossing your legs in relaxed company, hugging a friend.

Such adaptor activities are especially noticeable in small children, who have yet to be scolded or shamed into staying still. By the time we're adults, we've generally learned how to disguise these 'fidgets', but that doesn't mean they aren't still there – they often emerge as internalised gestures such as yawns, throat-clearing coughs, or other idiosyncratic tics.

What's worth noting is that adaptors aren't always indicative of anxiety or boredom. They may not even be negative. As we've already learned, to reliably interpret what a body is doing, we need to zoom out and observe the entire human in context. For example, it's been shown that 'fidget-ish' actions (such as foot-jiggling and doodling) sometimes aid us in maintaining focus – in attending to the task at hand – by dispersing our excess energy kinetically. A neat bit of body–mind adaptation!

<u>Illustrators</u> are our most creative gestures. They're unique, usually automatic, flourishes that appear to accompany speech. Unlike adaptors, we intentionally use them when communicating with others to add emphasis and value to our conversation. (We might enact a big, expansive gesture when we talk about something big, for example). As far as getting basic points across, they're not entirely essential. But layering pattern, feeling, and sometimes playfulness into our sentences is an effective part of our physical signature. In some people, they become so recognisable that they're almost like cliches: the politician's thumb press in their fist to stress the sincerity and urgency of their every point

or the fisherman using the distance between his hands to impress us with the size of their recent catch. Illustrators are a version of us 'talking with our hands' while speaking.

Emblems are the truest form of body, or gestural, language. Their meaning, good or bad, is always apparent. Emblems are a type of gesture that happens independently of speech. They convey a verbal meaning without words. Unlike illustrators, they can stand on their own without any speech at all.

A good example would be something we're all familiar with: the peace sign, a 'v' with index and middle finger with the palm facing outwards, thumb bent. You're probably also familiar with some, or many, emblems devoted to insults. Other emblems include air quotes, prayer hands, a power fist, and a raised middle finger.

Just as every culture has its verbal language, it develops culturally specific emblematic vocabularies and gestures. The emblematic vocabularies can also be gender specific. And what's true of national cultures can also be true of cultural organisations. One can find these emblematic symbolic gestural languages in sports teams, in the military, and even in dance companies. They tend to retain specificity to their origins – until they become full-blown cliches. We're generally pretty adept at detecting when an emblem has been appropriated by someone unable to convey it authentically. Think of older white guys coming up to you and doing a fist bump. You cringe because you instantly recognise it as an inauthentic gestural action.

Amanda Gorman's reading of her poem 'The Hill We Climb' at Joe Biden's inauguration in January 2021 was a fascinating example of gestural communication. It's a very controlled speech, performative rather than natural (no adaptors here). It's not spontaneous; everything is rehearsed. But she's able to communicate freshness through the fluidity of her actions.

What is dazzling about her performance is that Gorman passes through clichéd gestures, which we know and immediately recognise, to undermine them and cleverly subvert their meaning. At one point, very early on, when she talks about peace, she makes a peace sign but in both hands. It's a gentle, soft gesture that she kind of 'drifts' through, but wait – was that a peace sign or a gesture of 'inverted commas' – air quotes? Brilliant. While some of her gestures are culturally specific, others are culturally specific to her.

Her performance is a superb example of how it's possible to use a combination of illustrators and emblems to communicate with real physical eloquence. It's well worth spending time watching this speech online as a masterclass in gestural fluency. Or simply for the beauty of the poetry.

Dancing Gestures

Emblems are part of the fabric in the teaching of dance. They are used as communicative shorthand during training and rehearsal and for 'marking' (a process where dancers do not rehearse the movements fully but instead only 'draft' them, approximating the whole phrase with their hands). Dancers generally either know the emblems already or quickly start to understand them intuitively. In a ballet class, the teacher, usually an ex-dancer, will use gestures to let the class know what they want them to do. They speed up the sharing of a variation as the teacher replaces legs with fingers (or arms) to indicate battement, tendu, or glissade. So, for instance, they might say: 'Okay, today we're going to do battements', which they will illustrate by making a kicking leg gesture with their hands, followed by spinning their fingers to indicate a pirouette. Or they might teach a whole variation using their hands only: 'pirouette, battement, battement, glissade'.

Emblems also conserve energy during marking. This standard dance technique often involves substituting legs and other body parts for hand movements as dancers visualise their steps to secure them firmly in memory or so that they can physically think their role through while conjuring up new expressive touches. In classical ballet, emblems for teaching are standardised across companies. Within contemporary dance, companies generate their own.

Because gestures are so universally 'legible' – because every single body has a gigantic gestural vocabulary – they serve vital choreographic purposes, too. The most apparent use of gesture in theatre dance is its application as mime. Ballet and India's classical Bharatanatyam dance form, where bespoke hand motions are known as *mudras*, devised their extensive gesture lexicons centuries ago. Think of these as dance's equivalent of iconography in Renaissance painting, with specific hand actions to communicate the likes of 'peace', 'offering', 'embrace', 'betrothed', etc.

In ballet, these mimetic symbols can be archaic, necessitating study to 'translate' them, which has led many prominent companies to remove them entirely. Some are easier than others. Did you know that placing both hands on the heart is a gesture of 'love'? Perhaps a hand cupped to the ear signifies 'hear'. You might even surmise that crossed fists symbolise 'death'. However, you may not realise that ballet dancers tapping their forehead twice means 'princess' while tapping it three times signifies 'king' or 'queen'!

This need for translation in legacy dance is expected. As ballets are products of the times they are made, coded and loaded with the prevalent morals and modes of expression of their day, so is the gestural language embedded in their choreographic text. Through today's lens, the emblems of yesteryear may be considered outdated, uncommunicative and even offensive. Gestural language is a vital part of the authenticity of the original production. The dilemma of whether to adapt, edit, change or help audiences with the translation of a particular work's gestural meaning remains live for ballet companies worldwide. This re-evaluation of ballet text is an important one to consider and revisit.

Gestures change meaning in context and over time, much as societies and communities evolve. From an anthropological point of view, this is fascinating. Consider our twenty-first century obsession with technology (especially the technology we have in our hands) and how our gestural language and emblems modify as a result of using it.

Many of our life transactions are mediated through this tiny technology that fits in the palm of our hands. We talk, surf, pay, game, listen, photograph, and capture on these instruments, which feel like an augmentation of our hands – so central are they to our everyday embodied existence. (Who's lost their phone recently, and how did that feel? Like a limb lost.)

We no longer carry around chequebooks or pay for items in cash. We no longer pick up a phone receiver tethered to a wall and speak into its mouthpiece while keeping the earpiece tight to our ears. No longer are we taking a photo with our Leica (click click) nor leafing through the pages of an open book. Or at least, we might be, but Generation Z are certainly not – they may never even have seen a phone plugged into a wall. And thus, our conventional gestures for getting the bill, or for money, for 'call me on the phone' and 'please take my photo', and for reading are increasingly redundant – replaced by the new emblems: a card tap (pay), a flat hand next to my ear (phone me) and a scrolling gesture (with a single finger) for scanning the digital pages of a magazine. Our emblems are on the move – and some of our traditional ones are moving towards obsolescence. As our technology changes, so do our hand gestures and the culturally and generationally specific emblems we employ to speak nonverbally. Just like in ballet, gesturally, we are a product of our times.

At the opposite end of the gestural spectrum to ballet, where no translation is needed to decode the meaning of the gesture being performed, the postmodern American choreographers of the Judson Church espoused dance devised to be relatable to everybody, regardless of dance experience or prior 'ballet mime' knowledge. The use of 'pedestrian' movements was a crucial part of that form's language, and their emblems were basic human and recognisable actions like walking, falling, rolling, jumping, bending, and running.

In more abstract dance performances, recognisable gestures give the audience a sense of intimacy. In the final minutes of my abstract ballet *Infra* (The Royal Ballet, 2008), as Ed Watson's character struggles to piece his day together following a devastating bomb blast, he moves through 38 micro-gestures, performing scattered fragments of memory from the mundane – opening curtains; reaching up; headache; dirty shoes – to the horrifying: glass shard in a stomach; cradling a dying person's head; pulling debris out of his back. These serve as gestural annotations to the emotional 'touch and tenderness' female duet performed simultaneously on stage. In my abstract contemporary work, *Autobiography* (Company Wayne McGregor, 2017) the section 'Three Scenes' is composed of rotating duets of cast members repeating the same tale (a lover's row) using either naturalistic or heightened gestures and motions in varying combinations so that all viewers are drawn into the tiff.

Directing Gestures

Working as a movement director on plays, musicals, and films with incredible actors and directors is a significant and energising part of my career. I have had the great fortune to work with some of the best in their field – talented, highly skilled, and passionate individuals with whom we have collaborated to explore new territories of physical performance.

The job of a movement director, apart from choreographing any specific dances, movement sequences or intimate scenes in the work, is to prepare the actors (alongside the director) in their roles from a physical (and often emotional) perspective. I want to work with them to explore and investigate their character's movement behaviour and give them tools to act as a kind of embodied scaffolding. This process of preparation isn't about 'choreographing' every scene precisely but rather inviting the actor to extend their movement options at any given moment – in line with their character's story interactions – and thereby freeing them to improvise, play and invent within an evolved and character-'appropriate' vocabulary. The movement director also helps the actor recognise their own movement habits, the identifying features of their unique physical signature that may, if exposed too frequently in the role, upend or confuse their portrayal. This self-reflective process aids the actor in observing their rote way of moving and helps them discern when and where that may be useful to them and their character and when not.

The challenges and nature of experimentation in each project are totally different, even projects that might be part of the same broader franchise. For example, developing a movement language for Ezra Miller's tormented wizard character Credence in the *Fantastic Beasts* franchise is nothing like articulating the six-foot-four body double (on stilts) of Francis de la Tour's Madame Maxime in *Harry Potter*. Similarly, finding a violent physical expression of extreme mutilation in Sarah Kane's play *Cleansed* is not akin to mapping the female familial motion relationships in the National Theatre's production of *Dancing at Lughnasa*.

Yet, in all these examples, one of the foundational questions we always ask is how a character moves (or responds to movement) in various situations and, in particular, what dimensions and dynamics of gesture best serve the narratives they are telling.

In truth, preparing actors to examine their character's physicality and, therefore, unveil that character's intentional life is a complex process that mixes a range of physical (and emotional) intelligence and personal signature techniques. Gesture is a substantial one but not exclusive. Indeed, many of the competencies we have explored in other chapters come into play here: kinespheres, peri-personal space, attention, energetic baseline, spatial conditioning and so on – and that's without even mentioning scripts, words and meaning as motivators of action.

However, by looking at gesture in isolation, we can see how gestures contribute to a person's physical handwriting – be that in a character we are playing, someone we are meeting for the first time or ourselves. We can raise our awareness of our own personal communication signatures while learning to read others'.

Movement directing the film *Mary Queen of Scots* in 2018, in collaboration with director Josie Rourke, demanded choreographing masques, feral wedding dances, intimate love scenes and brutal assaults. Working with the entire cast, we set upon devising a total and varied physical world for the film, shaping the language of the courts and designing the delicate interpersonal relationships that unfolded between characters in the script.

At the movie's heart, Queen Mary and Queen Elizabeth, played by the exceptional Saoirse Ronan and Margot Robbie, alternated scenes, only once acting together in a poignant, albeit fictionalised, meeting of the royal cousins.

Spending time with actors of this calibre, who possess an extraordinary work ethic, is a phenomenal gift. Their genius, insight, and deep interrogation of their roles is inspiring.

Movement work is part of this alchemy, but I am fully aware that the actors always provide the assimilation, direction, and 'feel' for a part. I try to channel their instincts and open fresh avenues of expression with them, noticing what they are attending to physically at any given juncture in the script and offering them some further movement ammunition to play with.

Research in advance is vital at the beginning of any film process, especially when the historical context is unfamiliar. With *Mary Queen of Scots* set in the sixteenth century, the script provides the first forensic details of context, character, and situation. I initially pore over the script, highlighting the specific physical requirements in the writing, where there is explicit movement action and definite choreographed sequences. Getting an early sense of the character's journey through the film, logging it from a physical perspective, and visualising how the body alone might tell much of their intention, motivation and evolution in key scenes is essential to work before rehearsals begin. This is often preparation we undertake alongside conversations with the director where they articulate their vision for the story and how and in what style they want to tell it.

★ ★ ★

For *Mary Queen of Scots*, Josie wanted a fresh, direct and accessible acting style – a contemporary sensibility to permeate every storytelling aspect. Elaborate sixteenth-century costumes, in historically accurate styles (designed by the remarkable Alexandra Byrne), were made entirely of denim, and this playful counterpoint exemplified Josie's approach throughout. Dances could be anachronistic to the period, not actual sixteenth-century dances, but they had to be convincing in the ecology of the overall film. Similarly, the gestural language of the characters, especially of the two queens, needed to be invented. There was no attempt to recreate a literal physical impression of Mary and Elizabeth with gestures taken from paintings, text and historical description (although I did that research anyway!) but instead, an attempt to find an authentic and truthful language for them both that would be coherent in this imaginary world of the movie – an impressionistic rendering if you will.

In rehearsal, our task was to unlock, discover, and extend the individual movement behaviours of the two central characters through gesture (among other movement qualifiers: status relationships expressed in space, movement of royal ritual, etc.) – and the clues to achieving this were all in the script.

Mary and Elizabeth were written as kinds of opposites. They had had contrasting experiences growing up – where and how they lived led them to have inevitably diverse somatic histories. I came to read Mary's character's writing in a 'feral' way; she seemed visibly emotional, earthy and courageous. On the other hand, Elizabeth read as contained, breath held, emotionally distant, tight even.

These introductory qualities provided a perfect reference, a starting point to explore how, for example, adaptor gestures might influence their physical signatures for their respective parts.

The court Elizabeth presided over was radically different from Mary's (although both continually had to wrestle status from men's arms). The codes, behaviours and ritual conditioning could be encapsulated by their illustrator (while speaking) and emblem gestures (without speaking); these, too, would be contrasted. The court of Elizabeth and her movement vocabulary (including gestures) within it could subliminally feel linear, formal, rigidly structured and executed in a quicker staccato manner – that of Mary, rounder, circular, more recursive, openly conversational and executed more fluidly. And fascinatingly, from a gestural perspective, the scene between the two queens – the only scene where two women of equal power and vulnerability would meet – could encourage a new and distinct physical type of communication, placed further away from the performatively emblematic gestures necessary for maintaining court status and control. What if the language of queens in open dialogue was intimate, honest, and human-scale – full of the accidental adaptors and the unguarded gestures that they usually try to disguise, unmasking them and exposing their raw inner selves?

And that's how we started. We worked individually, Saoirse and Margot separately (never rehearsing together), and began to improvise, test, and try out various techniques in a range of scenarios to find and exploit their characters' physical voices scene by scene. Text and movement exercises, alongside experiments in expanding the palette of gestures, formed a solid base for Saoirse and Margot's character development in *Mary Queen of Scots*. But what happens when there are no words – when the scenes are entirely non-verbal?

Film is a visual medium; like dance, much of what is communicated on-screen relies on reading bodies in various contexts. Music plays a critical role in film too, either to underscore the verbal scenes with subtle acoustic support or to replace words entirely – providing a fully rendered score illustrating the emotional territory of the encounter, the interior landscape of the character, or an epic sonic image of scale and grandeur (think the mysterious opening of *Fargo*, the interior minimalism of *The Joker* or the swooping orchestrations of *The English Patient*). The relationship between music and movement, just as in dance, is key to any communication of meaning.

In *Mary Queen of Scots*, we diligently prepared many nonverbal scenes, the same way I would choreograph a dance. The 'text' or script here is an unfolding order of gestures and motion composed to generate specific visceral and emotional reactions in the audience. These gestures allow us, the viewers, to feel what is going on in the actors' imagination, empathise with their situation, and be effectively moved by their story.

Mary's final moments in the movie mirror the historical Mary's last moments of life. We see her private prayers and rituals with God while locked in her tower cell, her tender goodbye to her dearest ladies-in-waiting; we see her bravely walk her last steps towards death, momentarily passing through an outdoor courtyard where it snows, the flakes falling on her skin, she inhales. We see her face her adversaries as she enters the death chamber and we see her confidently mounting the stairs to her beheading platform. Her defiantly standing as her intimates pull back her black robe to reveal a scarlet red dress. We see her bend, her neck on the block, no blinking, her eyes wide open and... blackout.

Saoirse and I conceived this as an extended dance sequence. We improvised to music, sounds tempering the room and priming the atmosphere. We experimented in small sections – first the prayer ritual – and broke that down into many parts, inventing the gestural language (the emblems and the adaptors equally prophetic) to create a coherent miniature performance that could be repeated and varied. We moved on to investigating the next sequence (the tender goodbye) and the next until we had miniature dances for each key story moment.

We set them to music with a pulse, rhythm and 'feel' that best supported the sequences' development. We played them through as one long choreography over and over, and Saoirse would invent and change the phrase's physical inflexion, pitch, tone, and emphasis each time.

Saoirse is the most intuitive and exceptional physical actress I know. She is an expert at colouring in between the gestural lines, changing their meaning with a turn of the head, a flick of the eye and a hold of the breath. For this last scene, we rehearsed set choreography, devised gestures, and sequenced clear movement events. Still, it was the way Saoirse performed the actions, making live choices about their specificity and intonation and building on that knowledge, that made her execution of them so spontaneous, so truthful – in the moment.

The scene is a masterclass in communicating nonverbally. It teaches us that although we can usefully identify, categorise, and practise types of gestures to be at our communicative best, it is in their delivery that the most meaning value is extracted. Speaking with gestures is a dynamic act, a language enacted through time.

Saoirse is an actor who has refined her people-watching skills (including attending to gestures) to sharpen her portrayal of human experience in all its dimensions. She embodies this knowledge and turns it into practical techniques to convince us that the characters she is playing react and respond through their gestures as we know humans to do.

We can all sharpen our observation skills and transform them into techniques to refine our gestural repertoire. Using film as a tool to observe oneself is an excellent way to start, as the benefits of self-recording and analysis are substantial.

Ask a friend to film you during your conversation with another friend. Ensure they focus on your entire body from the front and obtain a clear image of you, including your hands. They should also record the audio.

Remember, no one likes to watch themselves on film, and it is easy to become distracted by your appearance and what you might want to change. However, the purpose of this exercise is analytical, and you need to apply a rigorous lens to it.

You should start by viewing the footage alone and without sound. I frequently employ this technique when watching a rehearsal in the theatre. I mute the audio to focus more effectively on the movement.

What strikes you first about your gestures here – what do you notice about how you move? Write your observations down. This is a *self-awareness* stage – watching yourself on camera can provide insights into what kinds of gestures you typically use (look back at the gesture categories above), how often you use them and how long you hold them for. This awareness can help you understand how you come across to others.

Next, examine the *identifying habits* within your gestural vocabulary: you may notice gestures you were previously unaware of, such as gesturing too quickly, too often, or not enough. You might also observe an excess of unconscious self-adaptors or object adaptors in your conversation, or you may find that your hands hardly ever move and that you are not very expressive. Note your observations down.

Now, repeat the exercise with the sound on. What changes? Are your gestures in sympathy with what you are saying? Do your movements feel stilted or abrupt? Are you gesturing as quickly as you speak? If you slow down your conversation, might that create space for more non-verbal communication and a wider gestural range? Note all of this down.

Ask your friends now to watch with you. Sound off. Then, sound on. What do they notice that you might have missed? How do they see and read your gestural language? Write their collective observations down.

You now have a brilliant page of focused gestural notes that will aid you in the next phase of this exercise.

By *setting goals*, you can identify areas for improvement and establish specific objectives for development, whether it involves enhancing your gestural fluency, experimenting with new types of gestures, or becoming more confident in communicating with your body.

Recording provides a means to track your progress over time. You can observe how your behaviour evolves as you focus on various aspects of your physical communication. This *feedback loop* is crucial for your continual growth. Lastly, *practice*: you can use recordings to rehearse different gesture scenarios – such as social interactions and wedding speeches – enhancing your preparedness and confidence in real situations.

10

We are Stories

Man with a pickaxe. Loose swings. Determined. Fit – arm muscles. Relentless. Hard labour. Sharpen head. Back to work. Face in shadow – pure upper body movement. Face in close-up. Concentration in the eyes. It is difficult to climb out of the shaft. Stilted rhythm. Hazardous. Unnaturally wide step. Slightly buckled legs. Extreme wind. Squatting position. Wild forces of nature. Arms protect the body, keeping it warm. Fists tightly across the chest. Drinking hot. Eyes singular. Hands in close-up. Chiselling. Hand with history – workmanlike. Spits on rock. Dynamite. Tired hands pick up the stick of dynamite and insert it into the wall. Hunched back. Materials. Light dynamite. Quicker exit up the mine shaft. Heave. Breath. Heaves materials. Heavy, can't do it. Explosion. Steadies himself. Looks in the shaft, shoulders tilted, fingers active, unsure, head nodding. Looks back into the shaft and carefully climbs back down the hole. The rung of the ladder snaps. Back hunches, hits the wall and falls to the bottom of the shaft. Air released. Black. Death rattle, extreme pain. Only the head moves and the body is immobilised. Numb. Stabbing chest, gulp. Battered body, sickled (obviously broken) leg, cradling the body and trying to pull himself out of the mine. Move. Stop. Fractured body position. Quick, sharp breaths. Grimacing. Scrabbles for stones, remains focussed on mission despite injury. Perseverance. One arm on the rope, one arm on the ladder – dead legs as he hoists himself out. Uses good leg as stabiliser. Pain. Rests. Tests strength. Goes again. Dragging himself across a stone floor. Heel pushes from good leg, the other limp. Feel the pain in the leg but ow more the agony in the back as the rocks slice his back. Elbow, hand, head. Moving faster, more agony. Cut to the mountain and journey ahead.

Growing up, I was always fascinated by the physical characters in silent films performed by Charlie Chaplin, Buster Keaton and Harold Lloyd, and silent movies like *Metropolis*, *City Lights*, and *Joan of Arc* remain some of my favourites today. More recently, *The Artist*, *Beau Travail*, and the extraordinary, wordless first fifteen minutes of Paul Thomas Anderson's *There Will Be Blood*, which I have just narrated, aptly demonstrate that there is often no need for words for us to understand.

One of the most prominent findings in psychology in the last half-century is that some facial expressions of emotion are universal, produced and recognised by everyone worldwide, regardless of race, culture, ethnicity, origin, or sex. Silent films can be enjoyed globally without translation. We cannot doubt Daniel Day-Lewis's character's determination, agony, and pain when we interpret his facial expressions in the scene described above. But this sequence is particularly intriguing because Lewis's face and eyes are in fact cast in shadow or occluded from view for much of it. The storytelling is all in his body: Lewis' movements, bodily positions and gestures.

Our ability to 'read' non-verbally seems hard-wired as we interpret meaning through facial expression, eye contact, gesture, posture, body language and context – where we are, how we dress and how we smell.

Take breath cues, for example. When you watch somebody speak, you can tell if they're gulping or swallowing. You can see it in their neck – emotion or anxiety manifesting physically as time intervals. Even if a person is standing still, you can sense their emotional state from the rhythm of their breathing or where their breathing 'sits' in their body – is it high in the chest or emanating from their stomach? You can tell immediately, for instance, if somebody is holding their breath and, therefore, interpret why that may be.

Therefore, we can instinctively feel when a crowd is about to charge, when a fight's about to break out, and when macho posturing turns into brutality. You can always sense, despite the smiles, that your best friend is not 'Fine!' Did you notice the moment your last relationship ended? Feel that irremediable break – the lack of eye contact, the 'safe' friends' touch, the negative space between you. Whether it's intuition, people skills, fast thinking, or empathy, it all comes down to physical intelligence.

Embodied Language

We are highly social animals, and reading others is extremely important for survival. We're born with an exceptional ability to get the measure of others: to anticipate their future actions by comparing their motions against previously witnessed behaviours, to pinpoint communicative anomalies and outliers – much like 'reading' avatars (this will make sense shortly!).

Without noticing what we're up to, we continuously spy fluctuations in other people's physical, psychic and emotional states, which we use to flesh out our mental representations of them. As we get to know someone, we daub this living portrait with our perceptions of and predictions about them and colour it further with the feelings and memories they rouse in us.

Even a passing introduction to a new colleague will arrive laden with assumptions about what this 'sort' of person might be like, with the stereotypes and biases we've accrued over a lifetime. Before a word is spoken, we scour for clues about relative power and value (however we ascribe these), well-being, personality type, and motivations. We assess potential fitness, contemplate whether their tattoos imply promiscuity, and gauge their trustworthiness from the movement in their faces.

Beyond helping us assess whether those nearby pose a threat or seem attractive, these usually unconscious processes – which are fed by our live sensory perceptions, mental preoccupations and goals, current emotional and physical states, and prior experiences – provide us with vital clues as to how best to communicate with our fellow humans.

Surprisingly, the science of 'embodied body language' – how we use our entire bodies in real-world settings to convey and understand messages – is a new area of research.[1] For decades, communication science focused on facial expressions and some gestures and poses, overlooking how we interpret or 'read' individuals in their environmental context. Yes, we primarily attend to faces and hands via perceptive systems that encode eyes, mouths, hand movements, and touchpoints separately. But simultaneously, we hear messages sent by a body's stance and motions and many additional factors. All of this – in concert – helps us interpret what the person before us is communicating, even if it's wildly at odds with the words they're saying. Whereas gesture was once the key component of our understanding of communication, this chapter explores its polyphonic nature.

Starting Young

Although we each have the innate capacity to observe human behaviour and interpret body language – to 'read' an individual's physical signature – these skills can be highly tuned with practice. The more we learn to observe, the more we can see. This repertoire of nonverbal behaviours starts in childhood. To help your child socialise, you must teach these skills, some consciously and some through unconscious imitation.

Consider gently touching your child's arm to let them know you're interested in and care about what they are saying or doing. While this might feel natural for some, I've observed many interactions where that connection is lacking – where the distance between parent and child is a physical, and thereby emotional, chasm. Turning to your child, bending down to their level and using lots of eye contact gives them a direct sense of engagement with you and instils nonverbal cueing. More generally, using your body language to understand their feelings is essential. Holding your children when they feel scared or sad sends a message to them more quickly than words ever can, assuring them that they are loved and protected. This type of encoded communication matters; surprisingly, it isn't always a natural part of building an intuitive connection with your children.

Games often feature excellent examples of non-verbal transactions. When I was growing up, my family played a lot of charades and card games, followed the leader, and guessed emotions through movement. These fun activities helped us develop receptivity to gestural differences and how to read them.

You will have spent your entire life learning nonverbal language. However, there is still time to improve and become more attentive to interpreting others. Observing movement is a skill that can be developed throughout life. The critical thing to remember is that communication involves the whole body.

ABBA Avatars

In 2020, I embarked on a thrilling, secretive mission to choreograph pop supergroup ABBA and create (with a superbly diverse team) digital counterparts of them, their 'Abba-tars', for their groundbreaking comeback show. The

Abba-tars would perform 'live' on stage, and replacing the original band would need to be as realistic and believable as the actual group. We needed to convince thousands of people every night that ABBA was back and that we were witnessing an iconic concert from the 1970s.

Unexpectedly, the ABBA *Voyage* project became a profound lesson in comprehensive body communication. It revealed that 'reading' live humans (accurately replicating human behaviours in a digital form) involves more than faces, gestures, and postures; it's about numerous subtle details in tandem. If even one element is amiss, the entire body seems unnatural. The human body is a sophisticated network of interconnected systems, making its digital replication exceedingly challenging.

Initially, we worked with the band, all now in their seventies, to help them regain their original performative expressivity and flair as they lip-synched in motion capture suits to (re)perform more than twenty old and new songs. This was a gift in terms of absorbing their unique physical signatures live. Even though our bodies change as we age, the fundamental 'youness', the way we move in the world, endures, as do your key identifiers.

Then, I painstakingly choreographed a set of younger human body doubles until they embodied those famed physical signatures as faithfully as possible. After twelve weeks of detailed and sharply focussed work – analysing, studying, mimicking, mapping, rehearsing, and replicating every single gesture and facial expression from our 1970s ABBA performance archive (and with a dash of contemporary artistic licence) – the doubles could perform the whole ABBA *Voyage* concert set – piano, guitar, vocals, dancing and all – immaculately in front of a small audience that included Benny and Björn! The body doubles did a fantastic job.

The next stage involved spending a month on the most extensive motion capture set-up I've ever seen. A team from legendary special-effects company Industrial Light & Magic then brought in motion capture (mocap) technology to record movements and apply them to a 3D model. It uses physical mocap suits, specialised cameras, and advanced software to create photorealistic animations for films, television, sports and even healthcare.

We translated the motion, facial expressions, and gestures of the ABBA body doubles into pure mathematical data. Simultaneously, an entire IMAG (image magnification) crew recorded every angle of the performance like a live concert. The result was a complete concert film and an accompanying dataset, which we utilised as our base. Ultimately, the high-fidelity avatars – ILM designed 3D models of the original ABBA to resemble them at age 25, with their bodies matching their actual proportions, and their faces realistically modelled – were presented live on stage and screened 15 metres high in close-up detail. Every minute feature was crucial. Indeed, the larger the face or body, the greater the chance the avatars would be less convincing overall, as our natural ability to spot anomalies, as discussed, is razor sharp.

The motion capture data from the shoot of the real ABBA's and the body doubles' performance was applied to the 3D models – the Abba-tars – resulting in dazzling perfection *and* imperfect chaos. Some of the maths translated seamlessly into beautifully articulated motion and physical phrases, astonishing the whole team with their accurate representation. But just as often, avatars levitated with their feet hovering above the floor, hands flapped like paddles, elbows twisted backwards, bones pierced through legs, and heads moved erratically. The notion that the maths of motion capture performances, translated into digital avatars in movies, is seamless was utterly shattered. Motion capture is an excellent technique for quickly collecting extensive physical data, processing it, and creating a (relatively) functional body. However, the human physical intelligence of animators, who later refined this raw data in extreme detail, makes the avatars believable.

This was the next phase of my task — forensically go through Frida, Agnetha, Benny, and Björn's performances frame by frame and assess the relative quality of the movement sequences (pedestrian and choreographed) and suggest improvements. Was this a convincing replication of the live performance? What action elements did the data capture well, and what did it leave out? How could we improve the avatar's motion performance and, in detail, fix the outliers? Was the meaning of the songs coming through in the avatar's version? What made the 'read' more convincing?

Back and forth, in a series of evolutions and alongside an animator team more than a thousand strong, we cleaned up the Abba-tars' physicality, gestures, weight, lip sync, expression and flow. The Abba-tars improved with each iteration, and soon, there was a considerably better, more fluid and integrated motion in all of them. However, we still discovered that the hands, the most expressive feature of any human body, needed to be animated manually. The avatars' heads, feet, eyes, facial expressions, skin, and breathing patterns also required improvement. Fixing one anomaly in one avatar would reveal others that needed resolution. We had to continually attend to the whole body, even if the fix was suggested for an isolated part – the isolated part in context mattered. While AI algorithms and machine learning played a crucial role, the expertise embodied in hand-animating each frame ensured that avatars became convincing substitutes for actual human beings. The human touch, the human eye, is irreplicable in this quest for authenticity.

Humans develop the ability to identify biological motion (the actions of a biological organism) by just four months of age. Following this point, we naturally enhance our people-reading skills through engagement in our communities. Such extensive exposure to reading others and their biological motion makes it nearly impossible to confuse non-biological motion with its counterpart.

We possess a primal, inherent 'uncanny valley' detector – an instinct that alerts us when we encounter any human-like figure that isn't genuinely human. This facility will have even more applications in the future, as we are faced with detecting increasingly sophisticated deepfakes or generated human-form content. While this innate faux-human alarm can promptly alert us to fake individuals, identifying which of the numerous subtle cues triggered this alarm may take minutes and sometimes even hours or days. Aside from the more common indicators like facial expressions, blink rates, misaligned eye contact, or unnatural joint and limb angles, multiple triggers invoke the 'uncanny valley' concept.

Breathing is key. If someone rushes around a stage or sings, their chest will rise and fall in a particular way. Singing forcefully will make neck veins and sinews stand out. Even though we may not realise it, if we can't see someone breathing, we won't perceive them as 'alive'. The ABBA avatar translation process emphasised how much we learn about others through the forces shown in their movements. Hair movement can convey more about our speed and rhythm than tapping feet. Air moving through flared silk trousers can reveal more about leg direction than the leg itself. We interpret bodies not just as bodies but through the forces acting on them. This helps us understand others' actions and predict their next moves. Without any effort, your body has gained an incredible grasp of physics.

ABBA Voyage is an extraordinary show. At the time of writing, the 2.5 million audience members have been blown away by this new form of entertainment, which uses digital avatars as persuasive live performers. However, recreating humans digitally is an extensive and delicate process. Even with the latest technology (which is constantly updated), human flesh and blood's ability to perceive and make sense of bodies as they communicate is unrivalled.

Our Bodies are Fiercely Territorial

I once received an invitation to conduct a workshop for CEOs on diversity of thinking at a well-known headhunter's office in London. On my way to the office, I entered an elevator where three men in suits were conversing. When I approached them, all three closed in, turned their backs on me, and colonised the lift buttons.

The message was clear. They wanted me to understand that I, standing there in my regular clothes, was not one of 'them'. I was the outsider; they were the insiders.

I didn't take it personally. This was not a new piece of 'choreography' for them. I imagine they would do the same thing any time they got in a lift – even (especially?) if it was with one of their employees. Maybe they'd pretend to be occupied (phone, book, nails), look elsewhere, and avoid the opportunity to engage. (Later on, around the table, I listened to one of the very same CEOs spout misogynist nonsense about the inability of women to design high-performance cars).

This memory made me reflect on how territorial our bodies can be and how effectively they can claim and control space. In some situations, it might manifest as someone creating a physical barrier around their work area, creating a personal bubble to avoid interacting with or acknowledging their colleagues. The message they're sending is clear: this space is mine! In such cases, the dynamics of a person's interactions change significantly when they enter that space. The most obvious, albeit comical, example of this territorial behaviour is waking up early on vacation to reserve a hotel deck chair with a towel.

Dancers are permanently defining the space around them. I notice in class and rehearsals that dancers are predisposed to always stand in the same place. The exact spot for a class, the same spot while creating, and the same spot for placing their water bottle and bag. And woe betide anyone who breaks the rules and stands where X typically stands. All as if there is some evolutionary advantage or safety in planting yourself exactly there – even spatial habits are ingrained.

This behaviour seems strange to me. I like to move around the space and choreograph from different places. One of the reasons we had bespoke curved ballet bars made at Studio Wayne McGregor was so that dancers could place themselves in different relationships with one another for class to encourage new interactive constellations – but honestly, the dancers, after the novelty wore off, tend not to use them, preferring their straight-lined counterparts and habitual spatial daily geography. I, too, have a specific place for my notebook, score, and water bottle. Rooms are encoded with meaning, and these meanings condition all social interaction.

You'd see many of the same dynamics if you were to find yourself in the bar of a theatre. People are determinedly staking their claim to the space around them. The bar at the Royal Opera House is an example. Step into the Paul Hamlyn Hall at the interval and you can see the groups and cliques forming instantly: the critics in one place, the bloggers, ballet fans and regulars each in others. I see one group occupying a particular cluster of tables and chairs. If anybody who's not in that group approaches, they use their physical presence to repel them.

I'm sure this will have happened to you, too. You will be at a party full of new people, and you'll notice that a group will have formed somewhere that colonises the space around them so aggressively that you can see no way to join them.

Building Bridges

You've likely seen the opposite situation as well: someone in a group notices that you might want to join and actively alters their position to create a welcoming bridge. They may adjust their body orientation towards you while conversing with another person, which makes you feel included. Their keen awareness of the social dynamics helps facilitate communication with you. By physically easing the tight formation of the group, they invite everyone to be more receptive.

Our ability to navigate social situations, especially those animated by status and power, is significantly influenced by our body language. It's about communicating effectively before a word has ever crossed our lips, and understanding this power can give us a sense of control in social dynamics.

Negotiating complex social situations like the theatre bar requires some physical problem-solving. But when you're confronted by a group that makes an impenetrable wall right in the middle of the room, it's good to know that there are techniques you can employ.

Here are a few strategies for social situations which combine the learning for interpreting bodies that this chapter has explored.

Observe people's faces, expressions or body positions if you aren't near enough; notice the closeness of the interactions; do you get a sense of sadness, shock or surprise? Who is engaged in private conversation? Who is dominating space? Physical mirroring is something else to search for. While matching and/or copying others' movements usually indicates fruitful dialogue – and is often a sign of empathy – it might, on further inspection, turn out to be a discreet mockery. If any interaction seems worthy of much closer analysis – for example, if you can't quite make out if a friend is being empathised with or bullied by the person they're speaking to – it's worth taking 'snapshots' at regular intervals to gain a sense of that relationship's progression through time.

How do you ascertain whether a group is open or closed? Open structures and groupings are welcoming, while closed forms and social circles seem more reserved. Keep an eye out for any signs of openness. Someone whose body is relaxed and open, rather than staring down with arms firmly crossed, is likely more approachable than anyone who seems wrapped up in themselves. Likewise, individuals who make grand, noticeable gestures and take up a good deal of space might initially seem less accessible than a quiet group that focuses inward. However, even a noisy crowd can appear inviting if there are clear entry points, like spaces made by people moving around, individuals standing at angles rather than in a tight lineup, or members of the group that meet your inquisitive look. These signs suggest you can approach with confidence.

Be sure to notice the whole picture – legs are more eloquent than you might anticipate. If someone is standing with their weight slightly backwards, jamming into their hamstrings, you'll know that they're more tense than someone with relaxed legs and whose knees appear to float. Paying attention to the direction of the feet, particularly when someone is sitting, can also be revealing. Feet that are unconsciously aimed towards the door can subtly hint at a person's desire to leave, a kind of unintentional signal that, even while engaged, they are mentally plotting their exit. This is a form of physical 'leakage' where the body's signals (if you notice them) undermine the presented intention of their owner. (We see this physical leakage when people lie, too!)

Someone leaning into the leg feels relaxed, and this sign of comfort indicates they are in no rush to leave. As far as they're concerned, their conversation will last a while. While it's nice to have conversations where both parties are comfortable, there are moments when you'd prefer to avoid being tied down in a lengthy chat. That's why, if I chat with someone I'd rather not be stuck with, I consciously try to prevent them from settling into a leaned position. I know that once they do, I'll be there for what feels like forever. However, if I enjoy the conversation and wish to continue, I might casually place my hands in my pockets. This stance is less formal than remaining fully upright, which involves more active engagement with the upper body and might suggest you're ready to move on soon.

Understanding that only some people are comfortable initiating conversation in a group setting, there are still effective ways to foster connections. If you feel out of place and eager to socialise more at a gathering, position yourself in a manner that invites interaction. By situating yourself within the space to be more accessible, you allow your physical presence to welcome more engagement. Strive for a posture that signals openness from every angle, demonstrating your readiness to engage through your eyes, facial expressions, and body language. This approach will make you appear more inviting for conversations rather than folding into yourself and waiting for someone to approach you out of sympathy.

If you ever feel the urge to withdraw due to societal discomfort – perhaps by tucking your hands away, avoiding eye contact, or appearing strangely stiff – it can be helpful to muster the strength to adopt the opposite demeanour. This means being expressive with your hands, maintaining natural eye contact, and allowing your body to move fluidly. (We discussed this in Chapter 8.) It's a harsh reality, but the more isolated or nervous someone looks, the more likely others are to keep their distance. Considering the interconnectedness of our actions and feelings, even a forced smile can make you appear more approachable and boost your mood.

The distance between you and another person is crucial to effectively communicating. Suppose you want someone to open up and engage expansively. In that case, maintaining a bit of space between you is often beneficial – not so far apart that it feels detached, but enough to create a comfortable boundary. If you are too close too soon, the interaction may feel rushed and insincere, resulting in a shorter, more guarded exchange.

When you establish your spatial limits, you can tailor them to fit any situation. To do this effectively, you must first recognise what those boundaries are. Applying techniques to recalibrate our understanding of peri-personal space and kinespheres – mentioned earlier in the book – alongside the 'reading bodies' tools discussed above will equip you with a valuable set of approaches for more effective communication.

Physical Scripts – One Frame at a Time

Attending to the *whole body* dramatically increases our perception. For example, watching live dancers work together in the studio allows me to gauge their commitment and collaboration by considering their entire form.

The problem is that it's challenging to parse multiple information streams simultaneously while staying fully present with those live bodies. This means I frequently need to break them down. Fortunately, dance and my experience provide a perspective for interpreting all interactions. We use a form of dance analysis to understand and read dance (or ABBA) by breaking it down into components. Once you learn this, it becomes second nature.

Dance analysis can be daunting, but understanding the elements of dance and human interaction generally revolves around a few fundamental principles: Body, Space, and Time. These principles can easily extend beyond dance to help perceive the physical context of any given moment.

While it may seem like we perceive the world as a continuous, unbroken stream, our visual cortex – the part of the brain that processes information from your eyes – actually oscillates rhythmically, delivering multiple 'snapshots' (somewhere between 25 to 75 frames per second) to the rest of your mind. We're naturally wired to interpret our surroundings one frame at a time. Start by taking a quick mental snapshot – an in-brain freeze-frame – capturing as much detail as possible and writing it down in memory as your 'physical script' for that moment.

Next, focus on the three distinct principles:

Body – the dancers (or other people) as individuals: gauge their age, gender, size, number, and role, all of which impact your perception of their current and potential movements. Alongside movement itself, they include body posture, gestures, verticality, extension, spatial dynamics, mirroring, and shapes.

Space – the visual setting: the space and props within it. What's the context? Is there a barre? Is it a room with many seats, or is everyone standing? Are there walls you can approach, or is it a circular room? What are the acoustics like? Is there loud music? Are people talking? Or are they listening to someone speak?

Time – how do these components interact at a specific moment, in and through time? Identify the relational dynamics within that snapshot. Or examine interactions over time and observe the relationship's evolution.

This framework aids group interactions by highlighting new indicators to watch rather than relying on familiar ones. These combined insights offer a robust understanding.

It's easy to misread intentions when looking at physical behaviours in isolation. In any given interaction, you should look at various elements cumulatively. Perhaps you are talking to somebody with darting eyes. That might tell you something, but not everything. If that person's darting eyes are combined with a tight body, they're in a very different emotional situation than somebody with darting eyes and a loose body. This information points toward a given intention. Naturally, you might be mistaken initially.

But when combined, it helps you refine your understanding. A single behaviour can imply many things. However, by linking it with others, you can narrow down the possibilities of what might truly be happening.

I take that snapshot to analyse a moment and focus on key elements: duration of eye contact, gaze direction, pose and posture, gesture, and touch. These elements are reliable indicators of motivation, intentions, and emotions.

Back to the dance studio: a dancer may hold eye contact with themselves in the mirror, their partner(s), or others in the room to evaluate who's observing them. The duration and nature of eye contact – whether it's swift, determined, empty, or unsure, whether it's looking or truly seeing – are strong indicators of the dancer's focus.

This applies to both performances and social interactions. When dancing or conversing, our eyes and facial expressions naturally reveal our intentions and emotions. Top film directors recognise this and tell their actors to concentrate on their scene intentions so their emotions surface through their expressions. Similarly, when we dance, we aim for our intentions and emotions to shine through our movements.

However, if someone is coasting, overwhelmed, or excessively confident to the point of appearing blasé, this will also be evident. Yet, before attributing a dancer's disengaged eyes to any cause, I will observe their overall demeanour for more clues. For instance, if their eyes dart around the studio, it could indicate confusion or the presence of a question rather than disinterest.

When dancers glance at themselves in the mirror a few times, they're likely to check their form – shape or position in space. Excessive mirror-checking, especially if it alters their body orientation away from partners, may suggest they're not engaging in the creative task or tuning into their body. Why else would they be looking externally?

To gently bring them back, I note the behaviour without an accusatory tone and address only that specific instance. I suggest they seem a little external. This same 'call it out' method works if someone appears more interested in what's happening over your shoulder or is distracted by their phone during the conversation. Instead of silently getting upset, mention it. The encounter will only be valuable if they're engaged anyway. This slight nudge can change the nature of your interaction.

The ability to predict others' movements at extreme velocity and maintain 360-degree awareness while focusing on one's partner, teammate, art form, or target is essential to elite performance in the culture and sporting arenas. Similarly, the skill of directing movement with just eyes is essential. During rehearsal, and even on stage, as dancers become physically tired, eye contact between partners increases – partly to stay on point and in sync when their bodies are under stress. Remarkably, this live eye-led negotiation settles the energetic value of the scene as together they then nuance their effort levels and emotional expressions. Without this connection, exhausted artists tend to overdo and misalign. Having reconnected eye-to-eye, the nature of the duet becomes more accessible to predict for all.

Recognising others' intentions is essential in cultural, sporting, or social interaction. Even a glance can provide valuable insights. But always look for additional cues: is the body tense or relaxed? Observe breathing patterns – is it high in the chest (shallow, anxious) or lower (natural, focused)?

Think about the synchrony between the people interacting. It's not just about the number of equal gestures; pay attention to their movements' range, placement, rhythm, and dynamics. Are they energetically aligned? Count the positive and negative physical cues. Do these indicate a productive conversation or a tense mismatch?

Did that collaboration or conversation start balanced and intimate – bodies close, eyes engaged – then deteriorate as one person took control and the other withdrew into defensiveness and non-cooperation? While eyes can indicate that an encounter is going awry, it's wiser to look deeper and longer at the whole person before drawing conclusions.

The more you view and engage with human physical behaviour across diverse activities, the more intel you will load into your incredible mirror cells. By attentively and actively focusing on how your body conveys your intentions and emotions, while simultaneously interpreting others' – especially through the medium of an expressive artistic practice such as dancing, acting, or music – your personal levels of emotional intelligence and kinaesthetic empathy will rise exponentially. The more we learn to interpret the stories our bodies tell, the better storytellers (communicators) we become.

11

We are Memory

Memory is a form of communication with oneself, acting as a complex internal dialogue that helps us navigate our experiences and understand our identities. It serves not only as a record of past experiences but also as a dynamic and ongoing conversation with oneself. It facilitates introspection, growth, and a deeper comprehension of ourselves and our relationships with the world.

The body is also capable of storing, recalling, and navigating memories essential to our self-development and understanding. We are a palimpsest of all our experiences: everything that has happened to us lives on in our bodies. Very often, this can be extremely useful. The body remembers and, on cue, can access these memories with incredible agility. Can you ever forget how to ride a bike? To swing high in the park, to perform a cartwheel? Ask a mature dancer to dance fragments of a work they danced when in school fifty years earlier; no problem. Ask a dancer to remember aspects of their hour-long improvisation; easy. Show a dancer a video of themselves dancing from last year and see how quickly they relearn the choreography.

At other times, our body's inability to forget can profoundly affect our mental and physical well-being. Not only does the body preserve our dancing, striking and cartwheeling action patterns and movement schemas, but it also stores away sensory and emotional markers of intense experiences. Emotions are embodied somatically, too.

Somatic Reminders

The body holds onto traumatic experiences through physical sensations and reactions. This unresolved trauma can lead to conditions such as chronic pain, fatigue, and autoimmune disorders.

These phenomena share the common trait of being part of our 'somatic memory'. Very often, we encounter them in the form of somatic markers: embodied imprints of past experiences that are rekindled when we feel something like that original state again, such as a maddening flush as we try on outfit number fifteen ahead of a date that stirs up memories of a similar flap before our last rendezvous.

Spaces and places can also trigger somatic markers; for example, we might feel noticeably buoyant when we find ourselves in a park we played in as children. Or, more disquietingly, we might become fretful as we pass a bus stop we were once robbed at.

When something major happens – good or bad – how we feel during that encounter gets recorded internally alongside our cognitive memories. This means that, in addition to being able to recount a play-by-play account of an experience using words, if we need to replay that experience on a deeper level or during a retelling, our somatic markers may be triggered.

As life is a mixed bag of ups and downs, some markers will enliven us or make us feel warm and fuzzy, while others may rile us up, even make us frightened. The point of somatic markers isn't to randomly change our mood. Our body and mind record these experiences because they were so impactful as to be notable and so that we can put this somatic knowledge to good use in future.

Somatic markers may trigger in the mind as our body tries to tell us something pertinent to our current situation, which might feel like intuition. This is why it can be wise to let a niggling fear or anxiety have its say in the light of day – if only so we can then assess if what our body is drawing our attention to is, in fact, useful. If it is, great! If not, we've rid ourselves of an unsettling sensation.

In this way, no matter the trigger, once we notice a discomfiting somatic sensation that doesn't have an apparent cause (for example, we're not hungry, and there's nothing amiss), instead of pushing it away – or fretting and so giving it control over how we feel and act – we can take it as a prompt to ask ourselves where that feeling might be coming from. Once we are able to figure that out, we can take heed of that message or dismiss it. And if that process itself gets a bit stressful, we can then deploy our breath, for example, to draw ourselves back to our energetic baseline or some other preferred state.

Another brilliant thing about recording our experiences somatically is that these sensations will have a physical location. Once we're adept at body-listening practices, such as meditative top-to-toe scans, or while working with a somatic practitioner, such as a masseuse or shiatsu therapist, we can pinpoint where that somatic memory lurks and try to release it.

For example, the lower back often stores a great deal of tension, and our hips tend to wind up housing a host of difficult emotions. For many, the tender abdominal region is where trauma resides, whereas our poor upper back and shoulders wind up carrying a lot of anxiety. Obviously, the precise location of these emotion-lodging spaces varies from body to body. We are, after all, all unique and process our experiences in different ways, but understanding that stresses, anxieties and fears are held in the body is the first step to dealing with these issues. Once we find a safe means of communication (solo or with any professional somatic practitioner), we can uncover where troublesome tensions and upsets linger and work to set them – and so ourselves – free once again.

The problem is that so many people either lack the opportunity or do not allow themselves to release. They constantly hold tension in their bodies, causing their emotional and physical scars to become increasingly entwined. As one part is bound, the surrounding muscles tighten, leading to overall atrophy in the area. It hardens like stone and becomes more brittle, restricting the person's ability to sit or stand.

Society has widely accepted the idea that we can discuss and treat our mental health challenges with a therapist. This approach is outstanding, as it has significantly improved the welfare of countless individuals. However, it's often overlooked that there are also physical equivalents.

It's easy to completely forget about the body (or at least to forget that the body and mind work holistically). Some people boycott their bodies altogether and for large stretches of time. That's an option, of course – choice. However, if you choose to work on it in a way that initiates that release process, it's like building a bridge. To me, it's a bit like how anti-anxiety drugs work: they give you cognitive space, a mental breather to view your concerns differently and examine your worrying from a new perspective.

That's why I believe in alternative practices like osteopathy, shiatsu or reiki. It's worth remembering that even once you begin that process of release, it's likely that you won't ever be able to remove the scarring entirely. However, you will be able to unlock some of those blocks and generate flow in your body in a profound way that can reconnect you with yourself.

We've all got our own scars; we've all got our own triggers. But if you can open up the body quietly and start to chip away at them, you're giving yourself that opening to experience a different kind of physical life.

The somatic detective work described above can also be a saviour in real time. Suppose we acknowledge that we are tense or anxious when encountering a stressful scenario. Then, we can direct our attention inside to feel where that sensation lurks and focus on releasing it.

Not so long ago, I had my teeth deep-cleaned at the dentist. This is quite an adventure because they anaesthetise your whole mouth and really dig. I could feel my body changing as I lay back in the chair. I am not usually anxious about visiting the dentist (in fact, I usually love going to the dentist), but I was holding my breath and felt my heartbeat pick up pace. This was weird: typically, an increased heart rate is coupled with breathing more quickly and panting.

It was clear that whatever I might have thought I felt about dental work in general, my body was, understandably, getting rather anxious – and signalling it dramatically. Keen to avoid spiralling into an even more anxious state, I drew my attention away from my mouth by performing a body scan.

I realised that I was bracing away from the chair, that my ribs and back were twisting with tension, apparently trying to evade the situation. More strangely still, as the scan travelled into my limbs, I found that my fear had 'jammed' my right elbow.

Odd, yes. But progress, nonetheless. This was absolutely a moment when I needed to work with my body and help it to give in – to let go. Now that I'd located the source of that tension, I could send my newly elongated breaths towards it while visualising soothing light shining through it until it gradually dissipated. By employing my breath and occupying my mind with the wish to dispel that grim sensation, I'd simultaneously decelerated my pulse and reset my nervous system.

Feeling calmer now, my mind wandered as I observed myself from a distance, and across time – an adolescent in the dentist's chair back in my hometown. The dentist then was a burly, brusque chap, working far too quickly in my mouth. I quite literally gagged. And there it was, the source of the tension in a memory triggered by sensation – fingers too large for my mouth, clumsily and inattentively moving in a rush, stretching my lips and nipping my tongue. To say it was uncomfortable would be an understatement; to describe it as careless is the takeaway. The first rule of auditioning a new dentist is to find one with slim fingers or, at the very least, assess their ability to work sensitively inside your mouth.

Double-Touch

Afterwards, I talked about tension, pain and desire with the dentist. She told me that she'd noticed that those patients who entered her surgery fearfully, expecting a painful experience, tended to experience more pain during their procedures than those who arrived at their appointments relatively relaxed. If you have a fear of pain and are very anxious about it, you register it more deeply.

She'd had men in her chair who were covered in tattoos – which would have caused them significant discomfort – but could not bear a needle in their mouth. Her conclusion has stayed with me: 'Because they desire it – because they really want it, the pain is reduced.'

There's a remarkable resemblance to dancers in this. Many of them have an unusually high pain threshold. Dancers know they can only attain incredible levels of fitness, physical control, coordination, and flexibility by pushing through a pain threshold that comes with achieving those things. Pain in service of something we desire feels worth it, and it may even be enjoyable.

This makes sense if one examines the inner workings of pain more closely. Pain is not merely an on/off siren. The term 'pain' covers a rapid-fire touch signal that alerts us that something is hurting right now so that we can immediately take rectifying action, such as removing the palm from the searing hot pot handle. It can also encompass a slower wave of discomfort and a touch sensation that inevitably follows the initial flare. Remember the last time you struck your funny bone? The initial pain is followed by a stinging sensation, almost like burning electricity. It's this latter nerve pain that feels worse, perhaps because you have mere milliseconds to realise it's about to hit.

This means that physical pain always strikes twice.[1] When we hurt our bodies, our vast, intricate touch perception system detects something's up via two different kinds of cells and sends two 'ouch' messages via two nerve pathways, each of which functions at a different speed. One is the fast track that sends the first flare-up to the brain, and the other is the pain wave that travels to the brain using our slow touch lane.

This slow-travelling element of the touch perception system is often called the C-tactile system. What separates our C-tactile system from our other touch perception streams is that its messages go directly into the brain's emotional touch centres. It's crafted to be both physical and emotional, a duality that grants it nigh on magical properties.

Earlier in the book, I explained how our sensory perception avenues, internal and external, continually send their signals. How streams of visual, audio, proprioceptive, touch and internal data (both perceptions and predictions) are beamed up to the brain all day long, and that it's then up to our minds (and bodies) to decide what's worth attending to and acting upon at any given moment. Having made a call about which is the most salient stream, we'll amplify it and hush up the others to properly focus on it.

This happens with our perceptions of pain, too. The pain we feel is always conditioned by the circumstances it inflicts on us. If, for example, we grab a hot pot handle, then after that rapid 'ow' flare has helped us swerve further harm by ensuring we let that pot go and get our hand under a cold tap, we will experience the C-tactile signal. Still, because this is an emotionally neutral injury – a silly accident, essentially – there's no primary reason for us to keep attending to it, so our pain signals can be quietened as we focus on mopping up sauce splashes.

But if the same mishap befalls us due to someone else's mistake (or if we take to beating ourselves up for it), then that burn remains emotionally pertinent, making it feel more intense – we'll be hurting emotionally and physically. You'll be concentrating both on the scale of the pain and your anger at that person who didn't warn you that the pan had been in the oven (or at yourself for forgetting); you might also be frightened that the damage will last. So, the pain feels incredibly relevant, and as a result, the signal is boosted.

In the case of my dentist's tattooed patient, his touch system would undoubtedly be shouting for their attention. Still, because he's familiar with the procedure he's undergoing and knows that it's improbable that anything terrible will happen and, more than that, has actively sought it out, he'll be able to shush those signals. He might feel discomfort but will be able to push it to one side and attend to something else, such as how great his sleeve is going to look once the scabs fall off.

By contrast, as he makes his way to that dental appointment, he'll be anticipating agony, and is naturally terrified. Because of this, he's exceptionally anxious, emotionally and psychologically. Even if he's not usually particularly aware of his body, of his interoception (internal perceptions), he'll likely become intensely interoceptively sensitive: he will notice his heart rate climbing, feel his skin dampen with perspiration, and begin to feel panicky. And that panicky state will make him feel yet more at risk, as if he can sense incipient doom.

Once in that dental chair, his pain perception system will be on red alert. Primed to be and feel hurt, any signs of discomfort or pain will be amped up and so grow yet more sensational in mind and body. The more you focus on pain and the emotion attached to it – the fear that it's only going to get worse – the more you'll feel it.

This tells us that we have far more control over pain and fear than we think. Our extraordinary capacity to work with and through pain, physical and emotional, is a tool you'll have (somatically) known and deployed for many years.

You Have Power Over Pain

How else do you think humans get through not just tattoo parlour visits but intensive courses of rehabilitation and physiotherapy, elite sports, dance and military training, pregnancy and labour? How else could we willingly – gladly, at times – work with and through levels of duress that terrify those lacking our unique desires and goals?

This is more than simply a matter of having a committed mindset. It's the ability to dismiss that pain as necessary discomfort and turn its signal down rather than concentrate on it and descend into a vortex of fear. Emotionally, that pain and discomfort will also be perceived as a sign of progress towards our desired goal ... it might even start to feel good. If we detach fear from pain, which is within our power, it'll lessen our discomfort sooner.

This dual nature of touch is one reason why, in recent years, more attention has been given to psychotherapy as a way of curing chronic pain in place of prescription medication. When there's no detectable physical injury, if we can calm that anxious-making 'something's terribly wrong' signal, there's a good chance the whole system will be positively affected.

Our C-tactile system also has further tricks to share. Pain is, in fact, just part of its remit. It's also the perceptive stream geared towards our 'reading' of hugs, caresses, massages, and all manner of affectionate and intimate skin contact. It has been shown that we enjoy skin-to-skin contact most at a speed of 1 to 10cm/second: if someone strokes our skin at 5cm/second – and treats us like a plush cat – the C-tactile system sends a luxuriant influx of oxytocin into our system. As well as feeling dreamy, this tide is likely to serve as a natural balm for any fear and pain we might be suffering. (C-tactile receptors can be found in high concentration in the middle of your upper back, a zone that's tough to self-stroke – perhaps nature's way of luring us into bonding with others.)

Even if no one is around to treat us like a favourite cat, we can still get in touch with our C-tactile system, soothing and revitalising our physical and emotional states through self-massage. This could mean something like a cosy self-hug or stroking our skin with a massage ball or soft-bristled brush. Whatever our chosen tool, it's primed a technique that we can call upon (along with breathwork) to ease our bodies and minds any time we like.

Taming Anxiety

I used to have a recurring nightmare. I am standing in the middle of a vast dance studio, pencil and notes in hand, totally frozen. A mummy stuck in time, surrounded by forty pairs of devouring eyes, all waiting for me to choreograph, to inspire. I cannot speak. I do not move. Lost. Then, in an instant, the cacophony begins – dancers offering a tide of suggestions, ideas, moves and improvisations. Forty individuals are pulling in their own sublime directions. Drowning in their creative noise, I see myself edge towards the back of the room – I watch on as chaos ensues. Blackout.

It can sometimes be hard to separate the process of creating and performing a dance from feelings of anxiety. No matter how much preparation or experience you have, the first days of a new creation always feel like the first days of your earliest works, and each live show brings a sense of high expectations, no matter how many times you've performed already. An underlying pressure suggests that you are only as good as your last performance, regardless of past accomplishments.

For dancers, the apprehension they might feel before a live show must be balanced against their ability to perform. Their nerves stem not just from the anticipation of dancing before an audience now but from the response they know will follow. That dopamine hit once it's over: the dancer's drug of choice.

As adrenaline thrusts into their bodies, they're increasingly pumped and edgy — when you watch them, it's clear that they can scarcely wait for curtain-up. They've trained for decades, prepared for years, and rehearsed for weeks and weeks on end. Their bodies are in peak form, their senses attuned and fully alive. They know their role intimately. They understand what they're doing, when and why, down to the tiniest gaze, breath and micro-gesture. At that moment, there's nothing more they desire than to move. They're in their ultimate state of preparedness.

However, as the dancers wait in the wings while the audience files in, they rarely seem anywhere near a place of zen-like focus. They fuss, shift, and practise things they've done a million times: pirouettes, jetés, pliés, and repeat stretches they've no need of — as if reassuring themselves their phenomenal body-machines still work. If the prosaic functions, all else will surely be fine. This excruciating hive of anxious energy persists until a hair-splitting beat before that curtain ascends. And then (usually) the gear moves into override, and the bodies fly.

What's distressing is that some performers *don't* manage to take flight. This extreme state of chemical unbalance, out-of-control nervousness, and profound anxiety hinder their ability to deliver, let alone excel in what they know they can do. They find even the basics are worryingly out of their grasp – each fundamental action feeling more distant than the last. On stage, they know; in the wings, they know; in the audience, they know, and this recursive circle of offences only makes matters worse as the dancer moves away from an in-body experience to a distant bird's eye view of their own unfolding disaster.

Dancers are inconsolable after performances like this, although every dancer has experienced this feeling to some extent in their career. But for some, the memory of these conditions, their fear response, and, over time, performance anxiety grow and become so overwhelming that they have to stop completely. Dancing on stage, live every night, is not for everyone.

This performance anxiety extends beyond the creative world. We have discussed the numerous individuals who experience a pathological resistance to and fear of public speaking, but for some, any social interaction beyond one-on-one settings induces significant stress. I have watched in pain as friends have stood in front of colleagues, family, and loved ones, struggling even to string their sentences together. It's not that they've stumbled over their words or that the words have somehow got in their way. Their physical state – their heightened nerves – has caused them to teeter. Their anxiety forces their natural centre of balance to be artificially high, resulting in fidgeting as they search for a resting point. Their mouth is dry, they smack their lips, their breathing is tight and shallow, and the moment they begin to speak – stumbling through – they rush and flounder.

The good news is that this doesn't need to be you. Performance anxiety in any field can be reduced with practice. Using strategies that prime your body to affect your mind or prime your mind to affect your body, you can control (or at least regulate) yourself to do better.

A sense of perspective helps – zooming out and seeing your smallness (this is a tremendous choreographic exercise). Humour, too (it is only a dance/presentation/speech, after all). You can employ your breath as your internal thermostat to dull the noise, opening an aperture to an inner stillness. You can also think about more long-term efforts. It is fundamentally important to build physical and mental resilience. Cultivating resilience gives you the energy to stand again when you fall. It's a central part of a dancer's training: to ensure that setbacks, injury or the constant daily grind don't derail them.

Of course, you can call on your physical intelligence. Physical exercise is a fundamental way of fighting the anxiety built into our systems, and you probably already employ it daily: running, dancing, jumping, skipping, hitting, boxing, or other intense cardiovascular physical activity is brilliant at dispersing surplus anxiety. A 2022 study on what are known as high-intensity interval training sessions proved that the more intense the activity, the more anxiety abated in participants' physicalities.

Somebody anxious ahead of giving a best-man speech at a wedding probably won't be able to start skipping around, and a nervous athlete will not want to waste energy ahead of a race. In both those instances, focused breathing would help manage their state of mind. However, they could also consider adaptors – self-oriented gestures such as hair-twirling, foot-tapping, or pen-clicking that we encountered in Chapter 9. These can aid us in focusing on whatever task we have at hand by serving as a form of self-regulation (engaging in repetitive, tactile activities can be soothing, and the sensory input from these actions can help ground you and bring you back to the present moment) and releasing nervous, or otherwise excess, energy through the body, thus calming the nervous system. Fidgeting, or what looks like fidgeting, really can help reduce anxiety. Everything from fidget spinners to stress balls to putty, even textured fabrics or objects, can offer the kind of soothing sensory input that helps relieve tension and stress.

Contemplating this might change our sense of what those dancers were doing backstage. So much of their behaviour – endless développes, re-checking their flawless make-up, muttering nonsensical phrases – might be read as signs of stress. For some, that might be true, but for others, it's precisely this repetitive, seemingly pointless behaviour that helps explain why, by the time they hit that stage, their anxiety will have vanished.

When paired with a total focus on their performance, these actions go beyond mere fidgeting; they fulfil the basic formula for ritual activity. What distinguishes ritual from routine or habit is its composition of a rigidly adhered-to sequence of actions, which do not directly accomplish anything, with a designated intention, which may be stated volubly or repeated internally. And we humans have been harnessing the physically intelligent powers of rituals to surpass ourselves for millennia.

Ritual Antics

Although recently re-popularised by 'gurus' prescribing corporate wellness initiatives, billions of people continue to perform their rituals daily in the guise of prayer and other sacred rites. Many of us will also have encountered pre-game rituals such as vibe-upping group chants, high fives, fist bumps and hugs while participating in team sports.

We'll likely have watched them play out on screen, too. The tennis player Rafael Nadal is a famed ritual actor. Anyone who has watched him on the court will have seen how, before making for the baseline ahead of a new service game, he'll lay his towel over the courtside barrier, tug its corners to smooth it, and check and re-check that it is symmetrical. Then, once he has begun to bounce the ball ahead of raising his racket to serve, he'll embark on a sequence of behaviour that seems imbued with almost religious significance: pant-yank, shirt-tug, hair-tuck, face-wipe, bounce, pant-yank, shirt-tug, hair-tuck, face-wipe, bounce; pant-yank, shirt-tug, hair-tuck, face-wipe...

These actions might appear ridiculous, but they're essential to his personal state of preparedness and a champion illustration of attention-priming, anxiety-dispelling ritual behaviour. No matter what else happens during any match, regardless of the score, despite extreme highs and lows – from serve to serve and service game to service game – Nadal's ritual plays out in precisely the same way.

Although rituals are so deeply embedded in our social and spiritual lives that they're generally just accepted as things humans do when in certain situations – such as answering a call to prayer or bonding with our teammates – it's only in the last few years that psychologists and neuroscientists have sought to determine precisely how and why they function so effectively.

According to the authors of a meta-study on ritual practices, rituals can be categorised into three broad groups based on their primary affect.

Collective prayer is a class of rituals that is, above all, focused on promoting the group's allegiance to their cause and each other, sustaining personal and interpersonal harmony. While praying, each participant's individual concerns are left behind as their attention is subsumed by the need to faithfully demonstrate a particular rite's actions in synchrony with the others while re-avowing shared beliefs. This strengthens communal faith and unity.

Another such class includes the pre-game rituals – solo, à la Nadal, or team-wide – which are deemed essential to cultivating and reaffirming goal focus. This type of ritual functions because by moving through an idiosyncratic action sequence – for example, checking and re-checking a sports kit or dance costume in the same way – such as a repeated assurance of preparedness calms an anxious mind and boosts self-confidence. And, as we know, physical actions diffuse excess tension. At the same time, intention-oriented mantras channel focal attention towards a desired outcome.

A third ritual style is conducted to dispel individual anxiety by drawing attention towards a higher power or goal. This type of rite, which we've yet to discuss, can include personal prayer or transpire in distinctly atheist breathwork, stretching, candle-lighting and mantra-repeating series. Privately, this process of mentally dedicating oneself to a greater cause or power while conducting parallel actions yanks attention out of worry-loops and invites anxious arousal to escape the body. By the end of the ritual, the practitioner will (likely) find themselves significantly clearer and calmer.

Whatever category a ritual belongs to, it functions as the result of the mind and body working in tandem. Both bottom-up (body into mind) and top-down (mind into body) priming processes help shift us into a desired state of presence and preparedness.

It's up to you how you do or don't choose to incorporate these activities into your life, whether that be a simple fidget or a more complex ritual. It's worth thinking mindfully about how any given action affects your anxiety levels and focus. We all will respond to different inputs differently, so testing out a range of tools and routines is worth doing. Be promiscuous about what you try! And, of course, very few things ever work in isolation: these actions are most effective when combined with deep breathing, mindfulness and regular physical activity.

12

We are Empathy

What if you could design your experiences to improve your communication with friends, family, or romantic partners, giving you new ways to generate intimacy? You would, wouldn't you? It would be time well spent.

And we do attempt this. But we default too often to language. We are culturally conditioned to put our trust in words to mend arguments or bring others closer to us. Even when faced with a physical message, such as a raised eyebrow, our first instinct is to respond with words: 'Are you okay?'

This puts both sides of the communication out of kilter. Especially if you say the wrong thing – which is easy, as words can be imprecise. You can entrench whatever it is that person has brought into the room. And when someone feels misunderstood or misread, then that's alienating.

Intuitively, we have a good sense of a situation's needs, but we don't always act upon it. Books and our upbringing tell us to say something, but sometimes, *doing* something is what the situation demands. While speaking can create more intimacy (it can equally create more distance), our nonverbal communication is often the most direct path to it. Nonverbal behaviours play a critical role in creating and sustaining intimate interactions with close friends and those you hold dear. They can also serve as a warning sign that something is going wrong.

Think of a once intimate relationship that is falling apart. The signs might emerge less in what is said and more in small (at first) physical microaggressions. There may be an almost imperceptible tension when one of the parties is touched by the other. Or, you might notice non-reciprocal body language, the absence of mirroring, or the fact that there's no generation of warmth and closeness – which is what intimacy is in a relationship.

Positive nonverbal cues are crucial: they manifest the deep connection that preserves and enriches intimate emotions and thoughts. This connection is maintained through the mere experience of these thoughts and feelings and nurtured through interaction. Although it is possible to feel intimacy by oneself, it truly flourishes and strengthens through mutual exchange. In a relationship, there's a continuous exchange of messages, back and forth, akin to a physical dialogue built on the foundation of earlier gestural cues. A wave, a hug, an embrace, a push, a hold, a tickle etc. Intimacy involves a combination of improvisation and gestural flow coming together to create a unique bond between individuals.

But unless a shared meaning emerges from this series of acts, you aren't authentically communicating; you're exposing yourself to misfires. The challenge lies in identifying the gestural anchors that convey the connection you wish to establish with someone else, whether it's a kiss when your partner returns home, or a moment to gaze into each other's eyes and truly see and be seen. If those cues are lacking or, worse, absent, intimacy may become an issue in your relationship.

These anchors can be highly prosaic. Consider the act of passing your partner a glass of wine: what does it signify if they don't express gratitude or make eye contact with you? Does it suggest an ease and familiarity with one another, or a developing neglect? What do you internalise from these moments of repetitive inattention? Do other more positive nonverbal acts even them out, or do they accumulate building tension? Are these moments careless oversights or signs of increasing estrangement?

Paying attention to your nonverbal interactions and the physical spaces between your words is undoubtedly beneficial. The starting point is noticing things in others and things in yourself physically. This creates intentional empathy with other bodies.

Kinaesthetic... What?

Dana Caspersen, a contemporary dancer, is a true virtuoso. Her performances in Ballet Frankfurt at Sadler's Wells were always a revelation, leaving me breathless. Her technical brilliance and expressive depth had a profound kinaesthetic effect on me, inspiring empathy. I felt what she felt. Her physical abandonment transported me. As I watched her, I felt free; I forgot where I was and was transported emotionally. Dana's performances serve as a reminder of how art, in this case, dance, can be a transformation, a transaction of energy between beings.

Watching gymnastics, fencing, speed skating, and wild horseback riding has given me a similar feeling. Oddly enough, for a few months in the early summer of 2020, observing the terrible acts of police brutality against the Black community in the US evoked the same response, or its ugly, negative sibling, visceral fear and pained fury. This was a prime example of social media's capability to arouse kinaesthetic empathy, not just for entertainment but also for driving social change – a significant aspect of its compelling appeal. It reminds us that experiencing the emotions of others physically can fundamentally transform us at a deep level. Watching the videos, we experienced the crowd's agitation as a key to grasping their emotional state and detecting the underlying tension ready to burst. We could sense their heightened energy and predict their forthcoming actions. The group dynamics spiked our arousal signals, chemistry, and physical awareness as we felt the closeness of danger around them. The precariousness of the situation stemmed not from spoken words but from the movements unfolding before us. We felt as they felt.

Kinaesthetic empathy is one of the most prized phenomena in all human experience and one of the most ingenious functions of our minds and bodies – an authentic powerhouse of physical intelligence. Whether creating dance for stage or screen, with humans or robots, I always attempt to elicit kinaesthetic empathy in the audience. Like many of the other incredible things about being alive – such as love or consciousness – it's a phenomenon that's tricky to capture in words. Perhaps that is because it happens automatically, like a firework display going off inside you. Yet it is something you'll have experienced many, many times.

When, how, and why you've felt it is determined by the range of your experiences. As we've noted, your entire movement history is stored in your body. Beneath all visible identity markers, you are a breathing, chatting, walking, waving, running, hugging, dancing, and living archive. Some basic actions, including some hand gestures (adaptors and emblems), are universal.

You'll have absorbed others from the people you've spent time with – including family, friends, teammates, collaborators, colleagues, and your wider community, which includes people you've watched on television. Nonverbal cues connect us all, making us feel understood and part of a more significant human experience.

The actions we learn and practice, such as kicking sequences, dances, bike riding, and other complex coordinations, are embedded in our bodies as action patterns and in our brains as movement schemas. This is also true for all our social moves.

Whenever you observe someone riding a bicycle, your eyes and brain immediately recognise it: 'there's a cyclist!' This reaction is automatic perception, something that everyone experiences. However, if you are a cyclist yourself, you'll understand and interpret the situation much more deeply without even needing to analyse it further consciously.

This is because, firstly, you'll get a felt sense of how that person is riding their bike: are they highly skilled or liable to topple any second? While watching them, you'll unconsciously compare their motions against those stored in your body and mind, as informed by your bike-riding action pattern and movement schema – your existing embodied knowledge. This means you've silently made a technical judgement call on their physical (or kinetic) performance.

Secondly, just by viewing them go past, you'll be able to tell if you like their style or not. And by deciding if you're into what they're doing, you're making an aesthetic call. Again, this will happen without you having to reason; your mind and body will keep working in case that information proves useful.

Finally, you'll be able to feel out (at least to some extent) if that cyclist is enjoying themselves or whether they appear nervous. Your emotional take on what that cyclist is experiencing won't be down to a happy/sad facial expression alone but based on a 'read' of their entire form. Yet that 'read' will be miles more nuanced if you've previously experienced that state for yourself. For example, if someone whips past at breakneck speed, sporting a gigantic grin, you'll note 'happy human'. But if you're also partial to bombing it about town, you'll feel some of that particular joy for yourself. You'll empathise with them. (This will be modified by how empathetic you are. Empathy is a personality trait, not a virtue. Some of us are born with loads of it. Some aren't, and are no less human for it.)

All this means that when you see someone performing an activity you've experienced, such as cycling about town, you won't just perceive what's happening in front of you. You'll also interpret their actions as informed by your biking experience, giving you a visceral understanding of their kinetic motions. You will make a judgement call on how they're riding that bike and decide if you like their style: an aesthetic call. And detect what they're likely feeling while riding. If you've experienced something similar, empathise with them.

In doing so, you will have experienced kinaesthetic empathy.

For those significantly experienced in an expressive art form, such as musicians, ballerinas, conductors, and actors, the experience of kinaesthetic empathy is even more transformative. That's because profound expertise in an expressive aesthetic form, which centres upon aesthetic ideals such as beauty and the communication of emotions, has been proven to increase artists' levels of empathy – including those who aren't born with keenly empathetic personalities. (Top tip: more empathy, more intimacy. Get to the theatre workshop!)

While most people can manage with a broad, functional emotional articulacy due to their ability to convey emotionally charged narratives through their bodies, elite dancers and other arts professionals, such as actors and classical musicians, consistently demonstrate a capacity for more nuanced emotional interpretations.

Dance education demands that students infuse their every action with intention and emotion. After several years of such training, professional dancers become phenomenally accomplished at composing, sending, receiving, and responding to eloquent embodied messages, which is what empathy is all about.

They must scrutinise how their movements look externally (what shape is the body making? Is its line as it should be?) and feel both proprioceptively (are the right muscles contracted? Is this balance point sustainable?) and effectively (goodness, my heart is racing).

Simultaneously, they stay aware of the aesthetic ideal they're trying to capture with their body. At the same time, they're continually receiving feedback on their expressivity – on the emotional and intentional journey they're performing – from both their dancing partners and the audience.

Little wonder dance has been posited as the ideal cross-training platform for emotionally intelligent, body-to-body communication. It can help you access ever-richer layers of experience and feel more kinaesthetic empathy; it doesn't just transform your body – it changes how you see the world.

Take a moment to consider any instances when kinaesthetic empathy has boosted your viewing experience of an art form or sport with which you've previously had experience.

For example, if you were (or are) a college gymnast, eminent conductor, school soccer legend, flute virtuoso, avid cricketer, acclaimed actor, Olympian swimmer, committed raver, insane roller-blader, retired ballerina, skilled carpenter, passionate firefighter, dedicated pilot, expert rock climber, amateur dancer, or professional surfer… then your experiences of watching your speciality being performed live (and reasonably well) on screen will be estimably richer and more complex than that of any fellow spectators who've never participated in that activity.

It might help to put this book down now, turn on a screen, and play a couple of videos. Pick one of an activity you're an expert in and have experienced – and then another of something you've never done and barely know. Then, compare those viewing experiences, concentrating on the actions (kinetic). When forming an opinion on whether you're into whatever it is that person's doing (aesthetic), try to get an emotional read (empathise).

What's likely to happen is that while watching your speciality, sensing out those three angles will be interesting (because of your expertise), satisfying (as using our knowledge, physical or mental, always is), transportive (as your body almost feels like it's now doing the thing you're watching – and you're emotionally engaged, too) and relatively easy because of all of the above.

When scoping out the second video, you'll probably find that it's entertaining (at first, at least, because it's novel, and novelty offers us dopamine hits), but hard to form a solid opinion of on either a technical or enjoyment level (because you don't have the expertise to draw upon). It will likely come across as emotionally simple and flat, as you don't know how it feels to experience that activity, so nuances of emotion on display will be hard to detect; although, as above, this depends on how empathetic you are and what you're viewing. It might feel dull after a while, because the dopamine's stopped flowing, and you don't have the knowledge to invest in what you're watching – yet.

Observing Actions: The Science Bit

But how do our bodies and minds trigger those internal fireworks? Much of it comes down to special mirror effects. Fundamentally, we understand other people's actions because unconscious representations of their motions play out internally within ourselves. But, when this occurs, it's not our movement schemas that light up, but special 'mirror neurons' in those same brain regions. These specialist cells aid us in interpreting others by reflecting simulations of observed actions in our mind's motor areas – just without activating the muscles we'd need to do them. In this way, as we view an action sequence, irrespective of whether we consciously 'think' about it, these mirror cells grant us a personalised proximate template of the action. Picture it as a millisecond-long shadow play in the brain.

This saves us a lot of energy; imagine if you could only understand an action by doing it! But it's also helpful because once we have this internal template, as well as understanding what that action is – say a wave, a 10/10 lay-up shot, or a pirouette – we can figure out if it's something we'd like to respond to (wave back), copy (try that shooting style on when next near a basketball hoop), or adapt (turn that pirouette into our gymnastic turning sequence).

Astoundingly, these mirror cells aren't just found in the mind's motor – as in kinetic, specific – areas. They've also been detected in further brain regions, including those responsible for social judgement, aesthetic appreciation, and value. And those brain regions we use to process emotions naturally collude to produce empathy. In their totality, the mind's entire interlinked collections of specialised mirror cells, which make kinaesthetic empathy possible in the first place, are known as the Action Observation Network.

As discussed earlier, we have perceptive systems attuned to body parts, such as those that fixate on hands and faces. However, the Action Observation Network has been shown to capture entire action sequences rather than individual moves.[1] This is another solid reason to analyse behaviour by 'reading' whole moving bodies rather than using solo poses or gazing at eyes alone.

The mirror cells in additional brain regions inform our impression of whether movements are being performed as they 'should' and what value they have within that specialist field. These will underpin our aesthetic opinion of those moves: that is, they enable us to decide if we like them. Meanwhile, we'll get a standard 'read' on how the person doing the activity feels. This emotional response will also be fed by our current embodied mood and any emotions roused by the actions we're watching.

Naturally, our emotional responses will be more sophisticated if we have experience of what the person we're watching is doing. Yet even without it, the more emotionally intelligent and empathetic we are, the richer our understanding of that human experience will be, purely because the mirror cells in our emotional brain centres will feed into our 'read'.

When bodies 'talk' to each other through kinaesthetic transfer, it can create fantastic energy. We all have levels of physical empathy that we do not use or are simply unaware of. Some of us, like musician Thom Yorke, are experts in transferring physical empathy without even realising it.

Hijacking Motion

Thom and director Garth Jennings invited me to choreograph the music video for Radiohead's 'Lotus Flower' single. Thom was initially nervous about creating a dance video and following set choreography but he was willing to try, like the experimenter he is. I wasn't concerned – I had seen Thom perform live on stage, and he is a stage animal, highly physical and naturally an accomplished mover. I asked Thom, when he performed, what he was thinking about, and whether he let his body respond to the music, to its beat and rhythm, or if he had pre-planned his actions. He gave a surprising answer. Thom said he was most inspired when he looked at the crowd and saw them moving and inhabiting his music. He would get a buzz from this direct, visceral feedback and begin to 'hijack' the specific movements and energy from individuals in the audience and repeat them back on stage. He would copy someone's dance in all its idiosyncrasies and reflect it back at them in his own way – with his own inflection and individuality. Then, he would look out again and steal another idea. This was Thom's private improvisational dance with his audience. Afterwards, he could not remember doing it or what he had done.

Thom's physical mirroring technique gave me a perfect way to choreograph the song. What if, instead of Thom learning four minutes of choreography rote, we broke down the song into sections and deployed his hijacking technique? We could film in shorter one-minute bursts. I could dance in front of him, off-camera, and he could riff off that – just like scanning the crowd for action and improvising with what he felt.

I've got a precious video of me dancing to Thom's track, grooving in front of him. He effortlessly captures my energy, moves, and choreographic ideas. You can see him noticing what I'm doing and reinterpreting it through his instrument, body, and moving style.

Likewise, as I watched him, I tried to give physical feedback and reflect on his dynamism and forms in a way that would inspire him. It was great fun. Notably, it highlighted Thom's tacit skill in catching the slightest motion and energy detail and translating it live: he's had a lot of practice from his concert life. Still, even so, his kinaesthetic empathy was off the charts. And because he was no longer focused on himself, worrying about remembering choreography and concentrating on the bodily information, the input he was getting from me, he was both relaxed and inventive. The video has had over 73 million views and always makes me smile when I watch it.

What exactly was Thom doing technically in this 'hijacking' dance? He was mirroring and matching. Conventionally, mirroring and matching are techniques used in body language to create a connection with others instantly. Mirroring involves replicating someone's gesture or posture precisely as it's presented: if my colleague is scratching their chin, I scratch mine; if they are nodding, I am too. Matching, on the other hand, involves observing the same gestures or poses but replicating them after a short delay (if my friend crosses their legs, I also cross my legs, but after a little time). However, mirroring and matching gestures and poses represents only a few of the available options.

As your body moves and creates various shapes, we can also replicate and synchronise timings, energies, intensities, momenta, dynamics, and emotions – we can even match the rate at which we blink! This means I can match your intensity with my intensity and your dynamics (the hard and soft of your movements, for example) with my dynamics.

In a more advanced version of this matching and mirroring, I can match your energy without necessarily adopting your physical posture (I could adopt your energetic bouncing motion without actually jumping, transferring that energetic cue to my arm instead), or I can align with your rhythm without emulating your emotional state (so I could replicate the same rhythm as yours, using the same movement, but instead of appearing anxious, I could look bored).

This technique, this game of mapping one component of action or body state, translated into a different dimension, and adapting and reflecting actions in various ways, is a crucial method we employ in the process of creating dance, too. Over time, dancers can evolve an instantaneous ability to 'crosstalk' in this way and develop total fluency in mapping one dimension of the body onto another.

Importantly, there is a substantial difference between mirroring and mimicking. Empathy in mirroring physical action can be used to build rapport: think of matching a gesture to your friends or swinging your jacket over your shoulder in imitation of your boss. Although, if it's inauthentic, you notice it immediately, which can be alienating.

One definition of mimicking might be building rapport with one community by excluding another. It happened to me at school. Some students would make gestures about me being effeminate; others would giggle. The kids mimicking hoped that they could reinforce their social standing by belittling me (as they saw it), and in some instances, their tactic worked, as mimicking can be highly destructive.

An empowering response is for the marginalised community to reclaim that physical gesture and expand its use. Once, a peer exaggerated and amplified their physical signature at school so flamboyantly and fabulously that any impression of them fell way below par. I eventually learnt to stand firm – to adopt a robust physical stance, legs slightly apart, shoulders wide, head held tall, radiating confidence from my sternum, quietening my breathing. It's a classic and effortless power pose!

This mirroring/matching technique can also help adjust social dynamics. I practise this inside and outside the studio when I notice someone feeling anxious at a particular moment. I reflect back to them their most recent physical move, effectively initiating a game of physical mirroring. This subtly signals that I've acknowledged their current emotional condition, which begins to put them at ease. Paradoxically, you can also draw people closer to you if you use a variation of this tactic – you can oppose their physical states. Let me explain.

What if someone exhibits aggressive energy towards you? You can mirror that hostile energy by reflecting the opposite back at them. For example, if somebody is confrontational and expresses it by standing animatedly forward, metaphorically *pushing* me back, instead of matching that force, I can sit back on my heels, demonstrating a relaxed state. I can slow down my breathing, and by not mirroring any aspect of their body state, I can de-escalate the intensity of their movements. I can drain them of their power.

Mirroring and matching are tools we can practise to generate kinaesthetic empathy, essential for fostering relationships and building intimacy. When we are drawn to someone, we instinctively mirror them, which helps build trust and show interest. Likewise, this strategy can reassure our significant others of our ongoing attraction and interest in them.

The profound connection from witnessing and empathetically interacting with one another's physical experiences surpasses merely understanding the words that describe these events. This capacity for emotional empathy is one of the most beautiful aspects of our humanity. We naturally react strongly when we see another person in distress, prompting an innate desire to comfort and support them. In dance, the goal is often to establish a kinaesthetic understanding with the audience, engaging them so that they may not feel the fear themselves but can sympathise with a performer perceived as scared.

Interestingly, this empathetic connection intensifies when an unexpected error or accident occurs on stage. Sensing a dancer's anxiety, tension or misstep, the audience collectively yearns for the performer to regain their balance and poise. This phenomenon almost feels like the audience conveys a sense of calm to the performer. I have often witnessed this where a dancer has been buoyed into excellent performance by the audience's empathetic transfer.

Equally, there's a unique joy in observing dancers perform complex manoeuvres from the classic ballet repertoire: the 32 fouettes from *Sleeping Beauty*, the powerful leaps in

Raymonda's ménage, the held balances of the Rose Adage. It's present too in moments like the dynamic headspins of breakdancers. As an audience, we experience a kinaesthetic thrill anticipating these virtuosic actions.

Watching dance can enhance our embodied empathy. The more time you spend observing and the more focused your attention, the more you can develop this ability. It might happen without you even being fully aware of it. Still, you're taking in a broad spectrum of emotions in a more personal and emotionally charged setting than any other. This is an experience you can't replicate by watching sports or observing people in your everyday life.

What if you could design your experiences to enhance communication in your closest relationships – whether with friends, family, or romantic partners – providing you with new ways to foster intimacy?

You would, wouldn't you?

Then investing time in dancing and watching dance will be time well spent.

PART THREE

Movement is Creativity

13

We are Fearless

Creativity is a reason to be and a way of being. It is ephemeral and concrete, a process and a practice, random and deliberate, giddy and profound. It serves as a free-play area and tool-bedecked workshop. It is a continually updating, acutely personal aggregate of skill sets, knowledge bases, techniques, experiences, and encounters, inspirations, insights, and intuitions – grounded in and emanating from body and mind – which may be applied in countless configurations.

Although its effects may seem alchemical, creativity is not some mysterious calling. As humans, we each possess the gift of invention – whether we access it most, some, or none of the time is fundamentally a choice. Imagination is a universal trait. Everybody is born with the ability to access and express unique conceptions; however, for those who are not artists by trade, artistic impetus and divergent thinking patterns are often suppressed as the convergently framed demands of daily life take over. Yet, just as we never lose the cognitive movement schemas, embodied epigenetic markers, and muscular cell volumes absorbed as we learn to swim or horse-ride, latent artistic propensities also reside within us. A creative person is any person.

Every creative achievement is a victory over fear. In creating that picture, song, or dance, we have overcome the anxieties that tell us our efforts are not good enough or that what we have attempted is impossible.

Fear can leave us blocked and bound. We might be afraid to offer ideas to others, lest they (and thus we) be shut down. Or we might fear thinking on our feet – this fear of improvisation in an artistic or professional setting mirrors the fear of improvisation itself. Tossed out of our comfort zone and in a panic, we will again shut ourselves down.

Throughout my professional career, I have acknowledged the inhibiting effects that various forms of fear and anxiety can have on me, both as an artist and as a person. I have cultivated strategies not to eliminate fears but to find workarounds to overcome them, thereby unleashing my own creativity and that of others. I have learned to make fear my friend.

We Need Fear

Fear isn't something to be scared of. It's neutral. Like all our emotions, when we remove the smoke and mirrors of our thoughts and feelings from the picture, fear simmers down to a simple chemical reaction: that flood of stress hormones, cortisol and adrenaline coursing through our body.

We need fear. For epochs, we have relied upon it to warn us of danger or jolt us into action. This superhero emotion signals imminent danger and kicks us into survival mode, priming us to fight or flee at lightning speed. It's an ancient alarm mechanism. Sure, hearing the alarm sound isn't pleasant. Stress hormones rip through your system: your heart races, your stomach churns, and your vision narrows. You imagine all kinds of nightmarish scenarios and painful past experiences, embodied traumas rearing up. It's a lot. Yet it'd be a terrible warning device if it didn't hijack our entire being, literally terrifying us into action. How secure would you feel if your fire alarm played lullabies?

Fear discomfits us by default. For solid evolutionary reasons, it distorts perception and impairs hearing and vision. Charged with preventing harm, it focuses on the defence and offence of survival. In this heightened state of aversion, our attention is fixed on the source of the threat, making it difficult to maintain an objective perspective, open awareness, or internal stillness.

Overcome by fear, we perceive others as hostile, agitated, detached, or constrained. Authentic connection and interpersonal trust become impossible, even though they may, in fact, be our best means of overcoming the state of terror we find ourselves in.

Granted, this is an extreme, and experiences of fear traverse a continuum. Each experience of fear is unique and influenced by our individual biology, emotions, and behaviour. Although our health will suffer for it, we can function from a fearful state for years or enjoy months without significant upset until a phobia is triggered and we lurch back towards the dread void.

For an emotion that persistently saves lives and nudges us towards our highest heights, fear has a terrible reputation. It gets shoved aside, repressed, and neglected instead of receiving the recognition it deserves for being one of our finest allies. Seriously, if we were in a real-life relationship with fear, it would have divorced us ages ago. This is simply because we're so hesitant to feel it that we force its messages into crags, feeding those demons and loading tension into our bodies. This blinds us to an invaluable trove of self-knowledge and stifles a wellspring of activating energy.

But it doesn't have to be this way! Simply acknowledging these routine, almost daily aspects of fear can diminish its power, and by exercising the techniques in this chapter, we can transform our relationship with fear into a potent, creative and well-being resource.

Flip the Script

With over thirty years of experience working with dancers, actors, schoolchildren, older adults, community organisations, and outreach programmes worldwide, I have observed that a consistently significant factor– whether among world-class ballerinas or pension-age dance beginners – is their relationship with fear and how it affects their creative abilities.

Although within each of them (and us) lies incredible strength, resilience, adaptability, and creativity, just waiting to be unlocked, when fear takes hold, it limits their possibilities. Fear in the body can hold us back. Over time, if we don't challenge our fears, the world can feel overwhelming and uncertain. This often leads to a comfort in routine, where our habits, thoughts, and interactions become repetitive. We may retreat into familiar patterns to avoid instability and fear of the unknown, crafting a safe but stagnant space.

But what if we flipped the script? What if instead of asking ourselves, 'can I?' we embraced the empowering thought of 'I can!' each time we faced a challenge? By shifting our mindset, we can dismantle fear's hold and open ourselves up to the endless possibilities that await. We can step beyond our comfort zones and explore our true potential. I learnt this mindset hack early, during my degree at Leeds, where I studied dance as a therapeutic activity for the elderly. As part of a practical course requirement, my peers and I developed a contemporary dance workshop for older adults, which included a performance and a collaborative dance-making session. While initially organising our tour, we faced resistance and concern from each care home director.

Although they welcomed the prospect of a dance *performance* for their residents (it was entertaining, fun, and distracting), they were apprehensive about the active workshop component. Each emphasised that their older adults preferred sitting and were reluctant to stand, let alone dance. Their de facto message was that their residents *couldn't*. This vivid example of others' restrictive perceptions of us, often as self-perpetuating and inhibiting as our own, was a challenge.

Nonetheless, after some persuasion, we embarked on our grand tour of Yorkshire's residential homes, performing in chintz-filled dining rooms to the tunes of the 1940s. We remained committed to doing more than entertaining; our goal was to alleviate movement anxieties and encourage the residents to reconnect with their bodies without fear, adopting an 'I can' approach. We always believed that they could.

As a hands-on experiment, we used lemons – our unexpected yet intentionally chosen device to engage our audience's bodies and imaginations! After settling the residents comfortably in their armchairs, each received their lemon. Knowing that laughter would relieve the tension and connect us, and with lemons in hand, we invited them to share their citrus stories with us: 'During the war, we washed our hair with lemons; I smelled like a fruit bowl… it attracted the wasps!' and 'It was great to finally have fresh lemons for pancake day!'

With the reminiscences and giggles fading and war-era songs playing softly in the background, we invited the residents to use their lemons for self-massage, drawing their attention inward. As their skin receptors sparked to life, they began to enjoy the sensations as their bodies reawakened. Their confidence grew as they used their massage device to explore parts of their bodies they hadn't attended to for some time – under the arm, along the shin, across their head, and even around their back.

We invited the group to massage their neighbour in the seat next to them – to reach across the armrest. Warm smiles filled the room as we turned up the music – and suddenly many of them were up on their feet and dancing! Others remained seated and danced their own perfect choreography with their chair as their stage. Now, we could begin the dance workshop – now everyone was in a can-do mindset: I can dance.

The care home directors and staff were amazed. Their so-called sedentary, motion-averse wards were filled with rhythm, flow, and stride for that session. We also sensed that the dancing participants had surprised themselves, transforming their movement doubts into a joyful, physical experience.

For older adults, unused to moving beyond the habitual actions necessitated by daily routines (and as we age, those routines themselves typically diminish), the commanding anxiety often centres around movement itself, and its attendant perceived potential for injury, pain, and strain. These limits are usually artificial. However, having embedded themselves in mind and body for decades, their binds are fiendishly tight. This defensive reticence to risk action in case of mishap resounded initially through every hesitation, 'ow' and nervy joint rub as our care home dancers began by demonstrating their anxieties via suspected ailments.

At first, profoundly disconnected from bodily awareness, the participants at our care home sessions could not interpret proprioceptive and tactile signals effectively, leading to a further lack of self-trust. They did not believe they could complete simple tasks (like rolling the lemon up and down their arm length) without their physicality letting them down. Once they moved to music and trusted their bodies to find a way, their anxieties dissipated as the sheer visceral pleasure of motion, of being back in their bodies, took over.

The 'I can' attitude is a powerful perspective that fosters self-belief and resilience. It encourages us to embrace challenges and pursue our goals with confidence. Nurturing this playful optimism empowers us to view obstacles as opportunities for growth rather than insurmountable barriers. Whether a super star ballerina trying a new dance vocabulary or a novice dancer in the class for the first time, the mindset barrier is the first to be dismantled. If you think you can, you can.

This technique has served me well, from creating my first ballet in an unfamiliar dance tradition to directing my first opera to simply starting a new work when I don't know where to begin. Over and over again, I have approached the unfamiliar with an open-hearted belief in the possible.

Combining this mindset with our capacity for unconventional thinking (lemons!) allows us to fully harness innovative problem-solving. In this example, how can we encourage older adults, who may feel somewhat disconnected from their bodies, to participate fully in a dance workshop? Especially when those around them thought it was impossible. Employing this type of creative, divergent thinking (moving from the 'can I?' to the 'I can') generates numerous potential solutions to any challenge and plays a vital role in overcoming our fears while fostering our creativity.

Thinking Divergently

Divergent thinkers approach situations with curiosity and flexibility, exploring multiple possibilities and perspectives. This combination nurtures ingenuity and creativity, inspiring us to take risks, learn from failures, and positively pursue our life goals.

★ ★ ★

Fear can significantly shape and restrict our divergent thinking, often acting as both a motivator and a barrier to creative expression. The fear of failure, criticism, or judgment can inhibit our willingness to explore unconventional ideas and take creative risks. This is particularly evident in environments where conformity is valued, as we may hesitate to share our unique perspectives because of the fear of negative repercussions. Consequently, our capacity for divergent thinking, which thrives on exploring multiple possibilities and open-mindedness, is severely limited. Artistic blocks frequently stem from a lack of faith in our creative abilities: from our fear of positing ideas lest they and we be shut down. As a result, we can become paralysed by fear.

However, fear can also catalyse divergent thinking when it motivates us to seek innovative solutions in the face of challenges. For example, the fear of an impending deadline (a premiere date!) might inspire us to brainstorm various creative ideas more urgently, leading to unexpected and original outcomes. In this instance, fear drives us to consider alternative approaches and to invent.

We all begin life with the creative instinct, the divergent mindset necessary to generate original and unexpected ideas. Unfortunately, due to the convergently aligned demands of most educational systems and professions, which consistently present questions with only one correct answer, right or wrong, as we mature, our nascent creative faculties are often abandoned.

We as a society seem to be ageing ourselves out of divergent thinking, stifling our natural creative prowess and transforming into less creative adults. Considering that we ideally use both convergent (related to judgement and evaluation) and divergent (linked to new ideas and imagination) cognitive processes to navigate life, a striking imbalance exists in how we prioritise and develop these abilities in both education and adulthood. This disparity is concerning because fostering creativity is vital for artistic innovation and innovation more generally, across various fields including technology, science, medicine, and design, making it crucial for success in our era of growing automation, innovation and AI.

In dance, a stark imbalance between convergent and divergent thinking approaches permeates our educational and professional contexts. In ballet schools, for example, while technical proficiencies (convergent thinking) are meticulously honed, there is a lack of comparable and parallel creative (divergent thinking) training. This gap is somewhat short-sighted, because the goal is to channel artistic talent into a creative career pathway. It's true that elite dancers in most forms must embody identical proficiencies and align beautifully in unison. However, that shouldn't negate individuality. Creativity versus reliability is not zero-sum.

Conversely, in contemporary dance, conformity is not nearly as prized as experimentation and daring in the pursuit of fresh artistic terrain. Here, creativity is fostered by promoting self-authorship, encouraging curiosity, and inviting questions – even about supposedly sacrosanct artists and works. Moreover, it is supported by direct training in critical thinking techniques, including observing, analysing, interpreting, reflecting, evaluating, inferring, explaining, solving problems, and making decisions in various formulations as the creative process evolves from the research phase to the final stage. (Although I must say that too often I have seen contemporary dancers in auditions with incredible imaginative and creative thinking skills, but a severe lack in their ability to execute even the most basic technical challenges, which is also far from ideal: there has to be some sort of middle way.)

It is no surprise, then, that international ballet companies have welcomed contemporary dance creators into their ranks to infuse new energy into their art form, often mining the diversity and expansiveness of their work and their willingness to embrace innovative creation methods. These artists have been trained to think outside the box, and it is through this divergent mode of working that invention, insight, and true inspiration emerge.

As a contemporary-trained choreographer, I have been invited to collaborate with some of the world's leading ballet companies for my ability to embrace and promote divergent thinking in ballet. This involves connecting unconventional collaborators and art forms, reinterpreting the ballet vocabulary, empowering dancers in the creative process, and encouraging a broader, more diverse audience to engage with the work. Cultivating my improvisational skills in these settings has been essential for nurturing flexibility and curiosity while exploring and implementing innovative ideas and concepts.

The concept of improvisation can leave many of us feeling anxious and self-conscious. Moreover, a fear of improvisation in any setting can indicate a broader fear of it. Therefore, our ability to improvise must start with accepting that it is possible and that we are capable of it. The freer we are in body and mind, the more effortless improvisation becomes.

Our ability to improvise involves conjuring countless original solutions to an unstructured problem, selecting the most promising ones while accepting that there is no definitive right or wrong. Suppose we habitually avoid exploring unexpected solutions to open-ended problems. In that case, we will find it increasingly difficult to guide ourselves out of trouble when life, in all its glorious chaos, deviates from our predictions. Tossed out of our comfort zone and panicked, we shut ourselves down.

What use is there for a football player who can't adapt to a change in play, or a camper faced with an uncooperative tent who is unready to create a makeshift shelter in the woods?

It's time for us, as a society, to accept the importance of adopting a divergent mindset for our creativity and collective survival. We must fearlessly unbind our educational systems from privileging the rational over the creative. Instead, we should welcome a new paradigm that celebrates and nurtures divergent thinking and improvisation as a mindset just as valuable as its opposite. Distinct processes of thinking that require equal nourishment.

Risky Business

Cultivating a willingness to enhance our curiosity and imagination through creative risk-taking involves exploring ideas that may seem uncertain or unconventional. This isn't always comfortable, as humans are culturally conditioned to seek certainties in our interpretations of anything: ourselves, others, films, records, and books. We are pattern-seekers by design – analysing, critiquing, and defining. The mind and body yearn for the safety of predictability and certainty, especially given the bias of our education systems. It's tricky to resist reason, stall disbelief, and navigate a way through, leaving a trail of open doors in our wake. What if we fail?

Our fear is malignant in this. Once it takes hold in our minds and bodies, it takes deliberate effort to disentangle what's real and merely imagined. It takes practice to ease out the knots in our physicality and quell nagging anxieties so that we may confidently navigate change and challenge.

The issue is that we're atrocious at assessing risk. This isn't entirely our fault. Humans are born with a negativity bias. From infancy, we create robust cognitive representations of negative experiences and perceptions. Negative information attracts attention and is weighted as more informative. We've evolved to dismiss a cascade of compliments in favour of stewing over a sole cutting remark. Even today, I vividly remember the scathing reviews over the celebratory ones – word for word!

This can make living in today's fear-hyping climate debilitatingly toxic. Fear is unparalleled in its ability to steal attention. We're steeped in ominous news. Social media is designed to evoke strong emotional responses. Many products are advertised as protection against decline and mortality. As we all learned during lockdown, living within a cortisol-sodden body amid a scared, aversive community only escalates stress.

We're overwhelmed attentionally, too. There is too much to be done at once, tended to, scrolled through to give us much thinking time – breathing space – to carefully deliberate provisional risk, to unpack each fear-spike. It takes perseverance to maintain perspective, cultivate presence and calm, trust others and ourselves, and stay open and unbound.

Thankfully, the collaborative mind-body approach we are advocating in this book focuses on redirecting attention from triggers, rebalancing hormones, and releasing physical tension without being overly intense. It might involve spending intimate time with loved ones, discovering wonder in a new awareness of nature, connecting with dance, art or music, engaging in deep breathing, participating in and enjoying physical adventures – each offering a means to reduce excessive fear and anxiety. Nonetheless, we are creatures of habit. We can likely engage in all these activities without straying from routine and relatively risk-free. But what if we want not just to alleviate but also to eliminate and eject unnecessary fears and anxieties? To nurture resilience, boost confidence and motivation, expand our horizons, generate endless possibilities, and fully unbind ourselves so we can act freely and improvise spontaneously?

An important step is to let go of doubt, shame, and embarrassment in the face of failure. Like in life, failure is an essential part of any creative process and a key to learning and growth. We should embrace failure and find comfort in it.

I have struggled with this concept in my practice, not only by attempting to compete with past successes but also by looking outward for artistic validation. Today, I can focus on my work as a continuum rather than a series of productions or products and have developed a way of thinking about my work that embraces the inevitable failure that comes from risk-taking in a creative process. Accepting that our work in the Studio is never truly 'finished' – that the choreographic process continuously evolves through failure upon failure – has been a great leveller and has encouraged me to judge my own work on my own terms.

Ideas surface and disappear, some to be caught and run with for a while, others missed only to re-emerge days, even years, later. A constantly shifting sand, where each rumination colours the next; multiple ideas flock and diverge, dissolve and ignite. With umpteen means to tackle every dance or process step, steadfastly sticking to the safest route only hobbles invention. You cannot eradicate artistic risk and other kinds of uncertainty from the creative process by obstinate planning, brick by brick. It's not just that proceeding like this gets tedious quickly; it's that the best ideas arrive glinting with dares and queries, not definitive statements.

Recognising that choices are not set in stone – that they can be reconsidered – is a liberating experience! This discipline encourages further risk-taking and helps us embrace failure. We can then move forward to explore the next question or challenge, continually engaging in the playful essence of creativity. This empowers us to explore and experiment without succumbing to anxiety. It serves as a reminder to artists, creators, and makers, and to everyone that any choice is always temporary and reversible – a means to an uncertain end.

Playtime

Like our physical signature, movement vocabulary, and communication style, our approach to risk – including a willingness to face challenges with curiosity and a tendency to persist rather than resist, freeze, or flee – results from a mix of nature and nurture: the biases and approach-or-avoid tendencies of those around us as we grow, as well as embodied practices and social interactions.

It's genetically imprinted and behaviourally conditioned. If, when we're little (and primed to learn through our bodies), we're coddled, banned from physical exploits in case of scrape, fall or impropriety, and repeatedly warned how dangerous our environment is, our outlook will become increasingly aversive. It's a shame, as this is much more damaging than a skinned shin or cracked rib. As you may remember, fearing falling only makes it more likely.

Our physical actions and reactions, tacitly conveyed, sculpt our offspring's self-conception, behaviour, and worldview, limiting (or expanding) their confidence, capacity to form healthy relationships, and so, ultimately, life choices. During the late teens and early twenties, any child may rebel, especially when hormonal patterns favour risky behaviours. But once seeded, a threat-based mentality is immensely tough to shake.

Whatever the impetus behind over-protective upbringing, the danger is that kids – of whatever background and physicality – may start to perceive themselves as fragile, injury-prone, and debilitated by design, which has a phenomenal impact on behaviour. The more fearful we are, the riskier our actions become. Not because we're hopelessly deficient, but because we lack foundational physical intelligence skill bases, such as complex coordination schemas, balance techniques, core stability, joint mobility, a felt sense of cardiovascular endurance at dynamic exertion levels, the embodied understanding it's worth serious discomfort to achieve a desired goal, for example. All of which makes activities trickier to get heads and bodies round – and the actual feeling of flesh in motion and accompanying interoceptive fluctuations feel alien, untrustworthy and scary.

Parents: please encourage your children to move, to expend their energy as they develop a range of skills by cross-training in diverse practices, to let them play freely, and to provide structured opportunities to train solo and with teammates, with instruments (of all kinds – bike, violin, trampoline, javelin, whatever), and without. Support them as they take pleasure in discovering the potential of their bodies and themselves. Feel secure that these endeavours further physical intelligence and prize their efforts here as highly as literacy, numeracy, and scientific endeavours.

These extracurricular pursuits do not need to become formal hobbies; this is not about accumulating grade certificates, winning tournaments, or holding captaincies for school applications or CVs – nor is it about alleviating your fear that they may not eventually succeed. It's clear that children are under too much pressure, far too early, from internalised fears over future prospects.

It doesn't matter if a chosen pastime appears productive or not; physical intelligence will progress if a child engages creatively with their body. The risks they take, the fears they face in playgrounds, playing fields, pitches, stages and rinks, and their resultant wins and losses, flights and falls provide invaluable teachings about assessing danger and overcoming challenges. These critical experiences instil courage, tenacity, improvisational skill and creativity, safeguarding them for decades.

We Humans

Fear can be a powerful ally, as can anxiety. The only way to excel is to dare to persist despite our fears and anxieties. It's all mindset. Heightened physical tension and cognitive strife, which can be excruciating when triggered by unforeseen threats, can fuel and even ignite brilliance. Even after decades of elite-level experience, show- or goal-induced anxiety unfailingly accompanies our biggest triumphs. Yet, with practice, performance anxiety in any field can be tamed through strategies which prime the body to affect the mind and vice versa, empowering us to control ourselves to do better. Centrally, breath serves as the internal thermometer and thermostat, enabling us to dispel disquiet so we may open an aperture to inner stillness, centre ourselves, and channel presence. In the long term, cultivating resilience is essential. Dance training helps hugely: where injuries, performance pressures and the ongoing grind attempt to derail us daily, resilience spurs us to rise again and again, no matter how brutal the fall.

No matter what obstacles we face, whether physical or psychological, the only way to overcome them is to face them head-on. If we minimise or ignore them because we're afraid of pain, hard work, rejection, failure, or death, we allow them to control us. To free ourselves from these obstacles, we must be willing to confront them directly.

Year upon year of working with hundreds of exceptional dancers while choreographing Company Wayne McGregor and the Royal Ballet has shown me that this holds as much for world-class professionals as much as it is for actors or novice movers. The higher we rise, the further we may soon plummet. And this is sensed only more keenly as we progress. We're only human, after all.

Demons

To keep fear from directing us, we need to accept it's part of the process – part of *us*. Artmaking, choreography, and dancing begin with the evolution of questions, rather than merely filling in answers. Indeed, artists strive to undo, unlearn, and uncover what they do not know so that they may start from that point. We need to move forward while asking the questions we are not yet able to answer.

Even when sought, this destabilising compound of uncertainty and reliance on the luck of the road makes for deep insecurity.

We all have demons which sit on our shoulders, needle, undermine, grow to an unimaginable scale, shaming us into not believing in ourselves or what we're trying to accomplish. Demons that damage and bully during angst-ridden nights of self-esteem-battering despair. We need to defend against their scaring us into settling for dead certs, coasting, binding our creativity until we produce works that we do not like: more demon fodder. But we mustn't dull them completely. They have their role, for they chide us into reaching further: never being satisfied with good enough. Indeed, our demons may be our most effective champions if we learn to speak their language.

Let's make fear our friend.

14

We are Invention

We're taught to think that creation is a siloed activity. That creativity occurs during art, design, music, writing, acting, and dance sessions, and that unless we're actors or dancers, the body is only involved in as much as our art form demands it, because the mind tackles the actual imaginative graft.

But hopefully we've now made enough progress in our journey towards physical literacy to know that the body must be considered creative in all art making: indeed, it is creative in any sensation, intuition or expression.

If the body is integral to our experiences and expressions, it follows that it is also responsible for generating our ideas and sources of inspiration. Our physical sensations, movements, and interactions with the world around us deeply influence our thoughts and creativity. The body is not merely a vessel for our consciousness; it actively participates in the process of thinking and feeling. Every idea we conceive and every spark of inspiration we experience is intricately woven into the fabric of our bodily experiences.

How can creation be a siloed activity when it's impossible to carve creative processes into distinct units of time or activity? Inspirations and insights take their own sweet time to arise, unfurl, distil and coalesce – sometimes only popping into mind or body decades after inception.

But, more importantly, we act creatively all the time – all of us, not just 'artists'. Every time we move non-habitually, in a way that's original to us, we're being creative. Every non-normative move you make is a creative act!

This includes playing astronauts in the garden with your nieces and nephews; stretching your agility via kinesphere games; manoeuvring furniture through doorways while moving house; walking backwards through corridors to test your proprioception; gesturing to get meaning across whenever words aren't enough; donning a VR headset and disappearing into *Beat Saber*; getting to grips with going downhill on crutches; calming anxious teens via TikTok dance-offs; revving up to give a big speech by pogoing about in the kitchen; dancing in the kitchen as if you can't help it; even having a crack at Twister at Christmas.

It doesn't matter if you haven't 'done art' since school; have no 'creative hobbies'; work purely with logic; function solely via cold hard reason (or so you imagine); are wary of 'creative types' and have been told countless times that you 'haven't a creative bone in your body'. Because you have a body – and because that body will have quietly been producing novel solutions to real-life issues, such as crossing icy streets with shopping in tow, or leaning to kayak with the children, for your entire life. I hate to break it to you, but – *you are creative!*

This chapter is focused on invention and creativity. Much of our previous understanding comes into play as we continue learning to conduct and perform our unique instrument – our incredible body – in increasingly sophisticated, expressive, and artistic ways.

We will develop the perception, visualisation, and projection skills we first encountered in the chapter on attention as we learn how to access, prime, and sustain flows of original ideas. We'll also examine the sorts of physical and mental creative habits that, while frequently handy, may conspire to stop new ideas or tip us into a 'no can do' spiral.

Creative Habits – In Mind

Many people recognise the crucial role of mental imagery in innovation and creativity. Along with many other art forms, dance relies heavily on imagery involving the body, movement, visual and auditory cues, and music. I have created pieces that focus on specific body parts – a dance of the hands, head, or back – and dances inspired by the work of painters Francis Bacon and Josef Albers. I have also responded to a wide range of music, from Bach to Mark Ronson to soundscapes of wolves howling, taking those soundscapes and adding imagery to them through dance. But imagery can also take on a more abstract nature. I have created dances that translate my feelings into colours, subsequently mapping those colours into moving shapes. I have also mined personal memories and used the 'feeling' of those moments to build improvisatory tasks. The influence of imagery on our creativity is vast and can be explored through various approaches.

★ ★ ★

Please imagine a tree. Hold it in your mind's eye and try to remember it.

For years, I noticed how remarkable it was that, when giving every Company Wayne McGregor dancer the exact same prompt for an improvisational task (a photo to respond to, a poem, a film, a feeling, or the cue to imagine a tree), each would react differently to it. Some would work on the shapes they saw; the emotional content inspired some, while others responded intuitively, using the stimulus as a vague instruction. And this diverse response would happen every time, even though the dancers had had similar training and were working within our shared movement vocabulary for any given piece.

I also kept clocking the dancers' repetitive blocks – the way that many got stuck in the same creative patterns over and over again. It seemed that when their initial burst of inspiration faded, their body and/or mind would slide into habit (repeating the same kinds of movements) rather than maintaining a flow of discoveries and fresh movement phrases. They weren't unaware of this: the dancers would sense the block, get frustrated by it, try and struggle out of it, becoming ever more fed up, only to find themselves completely stuck. An understandable side-step onto known terrain was repeatedly landing them on creative quicksand.

Such a situation is no fun for anyone and is an enemy of originality. Forcing anything, whether accomplishing difficult coordination, a complex task, or generating ideas, only decreases confidence. Wanting to find a way out of this recursive cycle, I investigated the mental and physical processes that facilitate the dance-making process. And I was lucky enough to do this in partnership with some of the world's finest scientists. While many fascinating revelations were to arise through these art–science collaborations, for now, all you need to know is that the very reason my dancers initially produced unique solutions to shared improvisational tasks was to provide the master solution to their collective block problem. But before we start unpacking those findings, please remember that tree.

It's an Image, Isn't It?

Although humans have an assortment of sensory pathways to play with, they frequently resort to the visual when creating mental imagery. This is to be expected, given that (for most of us) it's our dominant sense in terms of mental real estate, and visual media have increasingly dominated the planet's cultural environs, from general archetypes like film and photography to specific apps and games such as Instagram, TikTok, *Fortnite, Roblox,* Dall-E and the rest.

Therefore, when asked to imagine something non-specific, such as a tree, around 99 per cent of us imagine *looking* at it – as opposed to, say, feeling its bark on our palms or hearing the whistle of the wind through its leaves. If we think of a staircase, we tend to conjure a visual image, rather than listen to footsteps scampering down it or recall what sliding down a banister feels like. We also tend to visualise it from below rather than anywhere else on or around it. Similarly, when asked to mentally conjure a bridge, the overwhelming majority will describe their 'bridge' as viewed from the side over a river. Safe to say that our minds are absolutely stuffed with stock imagery!

Don't get disheartened, because this is neither unusual nor permanent: it's a standard creative thinking habit. And one that's not surprising either, for (as you've learned) our brains and bodies trend towards economy, and creating original material is much more energetically costly than flicking up a library image. On top of this, the sheer oversaturation of visuals that each of us contends with in our screen-tethered society has been shown to impoverish creativity.

A lack of free play in early years can also hinder creative capabilities in later life. These days, playthings mostly lead kids to *pretend* rather than to imagine (to fake bake or – yes, really – fake hoover), or to follow instructions: build Lego to numbered steps, colour in using numbered shades. From preschool to high school, toys and tools have a right or wrong way to be interacted with, while screens rule downtime. All of this leaves children with minimal chance for the active, imaginative flights (perhaps after an antsy dollop of boredom) that might magic up creatures, heroes, monsters, distant lands and Technicolor solar systems and originate their adventures!

As we know, the less we practice any physical or mental faculty, the greater the effort required to re-ignite it. No matter our age, the longer we neglect our physical and mental creative skill bases, the more demanding it will be to use them once we try to re-engage them. Much like our unthinking physical habits, the mental habits that we rely upon and deploy over a lifetime, such as how we approach imagination tasks, may later ditch us into the quicksand.

Again, it's never too late to change this — and simultaneously ourselves — because, much like the knack of bike riding or swimming, we never entirely lose our creative potential and skill bases. They lie low inside our minds and bodies, awaiting future activation and development.

At baseline, creativity is a collection of practicable behaviours, skills, and techniques imaginatively applied. So, through diligent and attentive practice, we can tune back into the internal sensations (such as proprioception, heart rate, and arousal state) that feed bodily awareness. First, we can relearn how to 'hear' and then summon, focus on, transform, compose, and convey our original inspirations, insights, and intuitions through our interconnected mind and body. So, what are we waiting for?

Back to that tree. Please load it up again. Is it distant or near? Are you standing by it, walking past it, sitting under it? Are you lazing in a hammock strung from its branches, or have you climbed it? Could you climb it? Is there foliage? Fruit? Is it edible? What does it taste like? Do you want to eat it? Does it scent its surroundings with resin, blossom, and rot? How old is it? How old are you? What season is it? Are you warm or chilly? Where is it? Atop the hill, on an island beach, in a neighbour's garden, as viewed from the window? Do its branches whisper, birds sing above it, squirrels scamper about it, or does it creak, storm-lashed? What does it make you feel? Joy? Sorrow? Excitement? Nostalgia? Is it a past or future tree? Has, will, or could it ever exist?

★ ★ ★

Did this game render you strangely tired? Unless you're used to this sort of play, it should. As mentioned above, to imagine anything wholly new, even something as basic as a tree, and hold it in mind while adding to it takes serious effort. Mental images are fleeting and continually updating them takes cognitive work.

What sense(s) do you usually *imagine along* with while reading fiction?

For example, do you primarily rely on your internal powers of vision (as we highlighted earlier) to depict scenes, also known as your mind's eye? This is common; however, there will still be a vast amount of variation. Maybe you internally 'hear' dialogue and other sound effects instead? Or 'feel' along with a story somatically, through a mix of touch, emotion and other bodily sensations?

Whatever your answer is – and importantly, there's no right or wrong here! – it reveals another creative mental habit. And one which, once again, all of us share. Most of us will have a favoured means of *imagining along* as we read, much like most of us have a favoured right or left hand.

As for dancers, including every single one in my company and at the Royal Ballet, those in every dance class and workshop Studio Wayne McGregor's ever hosted, those in every single dance school and hall, on every stage and dance floor on this earth, and all those who've only ever danced in their minds: all of them – that is, everybody – will also have a favoured way of responding creatively to any movement prompt they encounter, dictated in turn by how they cognitively process – how they think physically.

This (to take us back to the start of this section) is why, many years ago, all the dancers in my company would produce individual responses to the same prompts! They were (unknowingly) drawing upon their favoured way of solving a choreographic problem. In other words, they were reverting to trusted cognitive, creative habits. (The dancer who favoured shapes as against the ones that preferenced emotion or instinct.) That's also why, once those early ideas wore out, they couldn't sustain flights of originality. They would begin repeating themselves (physically or mentally), get frustrated, and then find themselves in a creative block. It was, in fact, their cognitive blocks getting in their way. For it's just not plausible to keep approaching the very same problem – mental, physical, or of any other kind – in precisely the same way, using the self-same tools, and maintaining a flow of original solutions!

This doesn't mean that any specific cognitive habit is bad or useless. Indeed, how could it be if it's served us beautifully for most of our imaginative lives? It is merely to say that in the very act of recognising that we have a favoured means of approaching creative questions – one that we can, with practice, opt not to use – we soon find we can access uncharted imaginative and physically expressive realms.

Let's return to the bridge I mentioned. What if you picture looking up at it from underneath in your mind's eye? Then, focusing on your mind's ear under that imagined bridge might bring to mind the sound properties associated with an echo. From there, connections can be made to ideas and emotions related to darkness or dampness, while moving back to the mind's eye might conjure the geometric properties of girders or brickwork. Imagining walking under the bridge might evoke an upright posture, but making it a low bridge over a small stream might lead to exploring the sensation of a stooped posture. This new body shape might draw your attention to the arc of your spine and the muscle tension in your legs. In this way, you have shifted the focus of attention across the entire landscape of the mind, including not only the mind's eye and the mind's ear but also the body and the deep schema. From a single starting point, each shift of attention brings to mind new properties that might be translated into movement material or something useful to you in that moment of experience.

Since discovering and learning how to work with mental images that can be visual, sonic, or kinaesthetic (or a mixture of all three), thanks to expert guidance from cognitive neuroscientist Dr Phil Barnard and arts and science researcher Professor Scott deLahunta, we often begin educational creative movement sessions – regardless of the attendees' age, ability, or experience – by helping participants identify their preferred sensory system for creating mental images. The instinctive imagery they bring to mind (whether visual, sonic, or somatic) is then expanded upon, and we guide them to explore that imagery by testing the alternatives. Just asking people the question unfailingly triggers a slow-burn revolution in both their self-understanding and creative potential. Much like me and my dancers – and yourself until a few pages ago – most humans are unaware that they have creative (cognitive) thinking habits at all.

Mind and Movement: The Imagery Toolbox

At Studio Wayne McGregor, we have developed a comprehensive training methodology, and a set of choreographic thinking tools called Mind and Movement to encourage the development of original dance movements by boosting imagination skills. This workbox resource, available at www.waynemcgregor.com, concentrates on three main training areas:

i) exercises that introduce a new movement creation skill
ii) exercises that focus on mental imagery
iii) exercises that use mental imagery as a stimulus for creating movement, thereby combining exercises one and two

We have successfully integrated these Mind and Movement techniques into many of our dance-making processes, which have proven to be both productive and inspiring. The full five-lesson course, built on twelve principles, is engaging and well worth your time – please do explore it online. However, I want to quickly highlight one of the training areas mentioned earlier, *exercises that focus on mental imagery*, which have been especially revealing, not just for dance, but also in other fields. Animators, writers, coders, cooks, designers, and composers have all experimented with this mental imagery toolbox, reporting that they have pushed their cognitive boundaries and strengthened their brain–body connections.

★ ★ ★

With this Mind and Movement technique, we draw on a theory from cognitive science about how attention is directed within our whole mental landscape. This theory is based on the idea that we all share the ability to imagine some things in the mind alone. For example, you can imagine your favourite bag or song without actually seeing the bag or hearing the song. Imagining your bag takes place in one part of the mind, sometimes referred to as 'seeing in your mind's eye', and imagining the song takes place in another part, 'in your mind's ear'.

The thoughts and associated feelings you have about your favourite bag or song form a third part of your ability to imagine things in the mind alone. This is where the imagination draws on your life experiences and brings meaning to the images you create in your mind. This third part is crucial as it is the source of your intuitive decisions and is directly connected to your bodily state. You can also imagine movement without moving (move your finger, now imagine moving your finger) and, for example, the sensation of rough sandpaper on your skin; these kinaesthetic and sensory images support these three parts. Using the focus of your attention, you can move around these different parts of the imagination and combine them in different ways.

Experiencing the possibilities of creating images and moving attention in the mind is the foundation for building skills in imagery. This is something you can easily practise:

Use your **mind's eye** to imagine the first image of a hat – any hat – that comes to you.

> Can you change the kind of hat it is?
>
> Can you make it a specific hat that belongs to you?
>
> Can you make the hat bigger or smaller?
>
> Choose an object in the space around you right now – any object – and add the hat to it.
>
> Can you imagine the hat as very big and the object as small?
>
> Can you put hat and object together in another scene?
>
> Can you imagine a sound landscape for the scene?
>
> Can you move your attention to your posture as you are creating the scene?

Use your **mind's ear** to think of something you know well that has a sound associated with it. Close your eyes and imagine that sound.

> Can you imagine the sound very close to you and then move it far away from you? Does the sound get louder or softer?
>
> Can you imagine the sound moving to the left or the right or undulating (going up and down)?
>
> Can you imagine the sound changing without moving it?
>
> Can you imagine a second sound in the background or another part of the space?

Can you imagine the sound in another scene?

Can you imagine a word that describes the sound (for example, 'ring, ring' or 'clomp, clomp')?

Can you take the word you just thought of and make a new object association?

This **imagine exercise** should be done standing; small movements in the body can be used to help the imagination.

Imagine you are holding a long, coiled spring.

Can you imagine the feeling of stretching the spring and releasing it?

Close your eyes and recall the image you created of expanding and releasing the spring.

Can you imagine the spring as much smaller and easier to expand?

Can you imagine it bigger?

Can you change the parts of the body being used at each end of the spring?

Can you change the orientation of the spring?

Can you be aware of a sense of muscle tension increasing with the expansion?

Can you imagine the texture of different surfaces of the spring? Can you change the texture in your mind, from metal to plastic, for example, or to furry?

★ ★ ★

WE ARE INVENTION

This introductory mental training marks the start of your Mind and Movement practice. It functions as imagery choreography. When you activate your imagination in this way — by asking questions about the form and structure of your internal images and creating new, improved versions — a vividness and richness in your perception gradually develop. This, in turn, enhances your body's ability to respond and be inventive.

When we create images in the mind's eye or ear, either static or dynamic, they do not have the detail of real sensations. However, these images encapsulate how sights, sounds and movements are organised and how they can change over time. These images are what we use in mental activity, whether we use our imagination for fun, solve problems, or mentally rehearse or create things. You can picture in the mind's eye or imagine how to rotate a sofa to get it through a door. You can also imagine how to execute a sequence of movements with a dance partner on the floor, but you would need your deep schema to come up with the idea that it could be better to try bringing the sofa in through a window or that the dance material might be more interesting if executed on a staircase, rather than on the floor. Similarly, you might use the concept of involving the mind's ear to imagine and rehearse a friend's phone number or some sound or music to accompany dance movements on the staircase.

It is worth emphasising that imagery is a very real form of mental work. It requires time and practice to develop and expand the skills, just like practising movement skills.

Developing these skills does not demand a deep understanding of cognitive or brain science or their theories. Very effective results can be achieved by following the simple, theory-based guidelines that we have incorporated into our exercises. Mind and Movement demonstrates that dance is not just about inventing movements but also relies heavily on various aspects of our mental landscape – a landscape that can be coloured, textured and primed in different ways.

15

We are Collaboration

Young dancers often ask me what it takes to 'become a choreographer', which allows me to explain the essence of my calling. You *live* dance-making. It's not a job or task to be executed. It's a deep connection with your body and mind and the physicality of the world inside and beyond you. Choreography is a philosophical relationship to and with your own life and others. As a choreographer, dance is the filter through which you experience the world. Dance guides the fullest you – the freest you!

When this realisation first dawned on me some decades ago, it served as an awakening. Choreography is a life of being perpetually attuned to yourself and others – to thinking through and with your body. Yet, somehow, this extraordinary way of being and acting – this awakened life – is totally misunderstood by most people.

The stereotypical image of 'the choreographer' is Debbie Allen's character from *Fame*. Standing at the front of the studio, teaching implausibly quickly, always to counts, as the dancers struggle to keep up. She stops, turns around: 5/6/7/8! And the dancers are expected to 'know' the phrase and perform it impeccably. Here we have the expert (Allen) and the novices (the dancers): empty vessels eager to be filled by this god-like giver of inspiration. Due to this fallacy, the beautiful, challenging and intensively collaborative dance-making adventure is largely considered – famed as – a one-way creative process. It is as if the choreographer *always* has every move pre-made and transfers them onto their cast to be danced precisely as shown. But this is only one way to choreograph, one form of what we can call a choreographic *process* – and there are many.

People First

I know many exceptionally talented artists who excel as solo creative technicians. Unfortunately, their limited interpersonal skills result in few opportunities to converse, collaborate, and exchange ideas, preventing them from reaching their full potential as creators. They lack effective communication skills – failing to empower, inspire, empathise, motivate, read, lead, listen, notice, share, open up, understand, and invite others into their experiences.

Recognising, rehearsing, and enhancing your interpersonal communication skills is essential for all forms of collaboration. Your interpersonal skills are the heartbeat of all collaborations, their vital signs. They facilitate exchange, nurture dialogue, and create both a process and a methodology to guide innovation. Collaborative individuals learn about how groups function, how they think and create together, how to maximise the potential of others, and how to allow space and time for each unique individual's voice to be heard.

Distributed Creativity

The beautiful byproduct, or perhaps really the ultimate reason for working with other humans, is that thinking becomes communal. Through a choreographic process, the dancers and I slowly become a singular creative organism – a generative swarm, unified by attention and intention, as we journey towards a shared ambition.

Once part of a group, whether we acknowledge this or not, key elements of each individual's attentive, perceptive, cognitive, emotive, and expressive faculties are modulated (or otherwise altered) according to those of the others. As previously covered, working together closely unfailingly affects each other physically, emotionally, and psychologically. We might even synchronise across our vital functions, for example, our pulses may align.

Humans' under-acknowledged excellence at cohering into a collective entity focused on a singular goal (creative, commercial, medical, social, or any other) is integral to the operations of team sports, art-making practices and performances, medical procedures, professional kitchens, and building sites.

While making and performing dance, artists constantly rely upon one another as 'things' to physically think with and through. Practising the art of embodied communication, we're frequently inspired by fellow dancers' intuitions and insights, which are communicated viscerally, immediately, via kinaesthetic empathy. Verbal discussions before, during, and after our sessions further assist us in finding novel solutions to creative quandaries.

This sort of active communal thinking is known as 'Distributed Cognition', a concept introduced to me by the marvellous cognitive scientist Dr David Kirsh, who spent many months with Company Wayne McGregor some years ago as we investigated the creative processes underpinning dance-making itself. After observing our methods for hundreds of hours, Dr Kirsh explained how the dancers and I were actually thinking together, distributing our cognition as we collectively invented, rehearsed, and performed new pieces.

The most extreme (and impressive) display of distributed cognition in action was the dancers' inter-reliant capacity to remember hours upon hours upon hours of material as it was created, developed, structured, and performed.

Company Wayne McGregor has no understudies – every dancer learns every role. At the Royal Ballet, while there may be understudies, members are simultaneously involved in multiple productions. For example, after rehearsing one piece in the early afternoon, they'll perform another that night and participate in the choreography of a third ballet the following day before rehearsing a different piece again and then performing in another that night.

This is only possible with pronounced joint attention and distributed memory skills. By remembering together, each dancer facilitates the others' brilliance.

At Company Wayne McGregor and in my productions for the Royal Ballet, movement vocabulary is created in packets by tasking and other methods, days, sometimes weeks, before it's 'fixed' — that is, locked down for performance by specific cast members at a particular time in the show order, and situated in terms of the music, lighting, set, and so on.

Until then, it needs to stay 'live' in all dancers' memories so that it can be re-accessed and worked into intermittently in later sessions. Additionally, at every stage of any work's development, while creating new adaptations for their phrases, the dancers also have to be able to instantaneously recall those phrases' prior iterations — that is, they need to constantly 'keep their originals'.

In this way, the choreographer's inspiration, idea, and intentions for each phrase are off-loaded onto the dancers, who capture and interpret them through their minds and bodies, solo or in groups, to create an artistic response.

Thankfully, every dancer's intention (the seed of their response, whether an idea, emotion, visual thought, somatic feeling, verbal quote, or internal or external imaginative projection) as they make a phrase helps them anchor it in their own embodied memory.

Yet, that phrase will then be further individually and collaboratively developed throughout the choreographic process. For example, it will be grafted, divided, embroidered, reversed, dismantled, reconfigured, and otherwise remade in many diverse ways as the original snippet morphs through variations, potentially winding up aeons away from its initial form.

Without every dancer and choreographer being able to rely upon their collaborators' minds and bodies to inspire, invent, develop, interrogate, and jointly remember all these hundreds of dance formulations – without distributing their cognition in all these ways – no dance piece would ever be made.

You have to admit, this is an incredible feat of creativity!

Creativity is Interactive

Selection also occurs throughout the choreographic process, from the first task explored through every developmental phase. Decisions and edits have been made, including structural, aesthetic, and expressive changes – the bare bones of composition.

As mentioned, while much of this improvisation and craft-led exploration is relatively free, it is framed by prior work on the creative direction of the piece, and by the initial parameters and possibilities embedded in that piece's music, set, lighting, staging, and costume design. Indeed, away from the studio, all these elements will have gone through their own extensive, physically intelligent design and making processes, led by my vision for that work.

Yet in the 'compositional' phase, I will fully sequence, adjust and finesse dance material according to how it functions once situated among these components. At this juncture, the judgement of movement material shifts from a distributed enterprise to an audience of one: myself as the choreographer.

The relationships between the music, lighting design, stage architecture, material or video imagery, bodies, and costumes are sharply focused during this period. As everything affects everything else, unforeseen artistic challenges and opportunities appear. Only this time, players from a panoply of practices congregate around our moving puzzle.

Ideally, these elements will have been delicately engineered and nuanced by the cusp of curtain rise to coalesce as intended to transfix and transport our audience through kinaesthetic empathy. But before we hit that stage, let's skip back a few months (or years) and hop outside the studio to spotlight how another unsung, innately physically intelligent skill base assists humans creatively.

Interactive cognition – actively thinking with and through anything: pencil, bow, tong, model, keyboard – aids us daily as we perceive, interpret, reason, predict, produce, communicate, act, react, and otherwise navigate our planet and its inhabitants.

With practice, our interactive behaviours such as sketching, typing, wood-chopping, stilt-walking, and piano-playing become coordinated enough to seem unthinking, habitual, as the instrument involved is subsumed into our body schema. During creative pursuits, 'interactive cognition' is fundamental to being able to progress ideas, as by externalising an iteration – for example, penning a stanza, cutting a pattern, tapping out chords, sculpting paper models – we have something external to our minds and bodies to help us assess our inspiring idea's viability and envision potential evolutions. As powerful as our imaginations are, everything we create will always be experienced within its intended context and environment. Bringing it into reality will best serve us in getting a clearer picture of its eventual effect and affectivity. This enables us to perceive it through an array of sensory streams, thoughtfully re-consider its attributes, and project provisional changes onto it.

This advanced style of physical thinking – the imagining of adaptations to an idea rendered as an object (or being), such as a drawn sketch, musical score, design pattern, or 3D graphic model, with the specific intention of changing it – is known as 'enactive projection'.

Every key collaborative partner – designers, artists, lighting directors, set builders, seamstresses – involved in dance production will have a means of moving their creations along this path.

As for the choreographer and dancers, although we may frequently use notebooks, notations and videos to capture aspects of our creations, we can only access and feel their true depths and evocations by thinking them through physically. So, as our bodies are medium and message both, our 'enactive projection' is often embodied. In 360 degrees, working with gravity, we can truly sense, feel, contemplate, and intuit any phrase's impacts and effects and so project appealing continuations onto our idea from within and without.

For example, instead of formally noting down potential next steps, we sketch our phrases with our bodies – a practice known as 'marking for self'. We may also riff off our fellow cast members' motions to test them out on our own bodies while making together. By deploying our bodies as fluid tools to trial and develop ideas in this fashion, we can fully attend to and adjust that physical idea's real-world possibilities and select and nuance them into our desired form.

The dynamic relationship among dancers, space, time, and energy that underpins every aspect of the processes described above embodies a form of collaborative flow. It often evokes a blissful sense of being lost in time, surrendering to the body, and being consumed by the dance.

16

We are Fluency

I'm drawn to ensōs, those Zen Buddhism circles meant to be created in a single, free-flowing brushstroke. The beauty of the final product is captivating, as is the process involved: sitting in meditation before ever touching the brush.

Sometimes, the circle you draw brings you satisfaction; at other times, it doesn't. But even experts who dedicate years or their entire lives to this practice never truly succeed in drawing the ideal circle with perfect, even brush strokes – and that is the point. Achieving complete mastery is an elusive goal; there will always be shortcomings. Yet the allure of attaining such perfection, where conscious effort and subconscious skill align, remains irresistibly appealing – the desire for it never fades.

Have you ever considered what fluency might mean when applied to something other than verbal language? If so, that's unusual. Although humans can be or become fluent in just about any activity, we tend to associate that condition with linguistic prowess alone, which is unfortunate. Everybody understands what linguistic fluency is – the capacity to speak eloquently. To spontaneously flow between grammars, styles and vocabularies. Switching vocal cadence, rhythm, tone and pitch harmoniously with live demands. To speak fluently is to communicate meaning, convey intention and express feeling with ease. Without fluency – no craic, no gags, no improv – your chat is dry. Without fluency, no innovation. A fluent speaker can bend rules, upend conventions, deviate mid-sentence without losing their audience.

Physical fluency is very like linguistic fluency. In physical fluency we feel alive and free, we soar beyond limits. We dissolve into presence, body into motion, pure flowing energy. Awareness, sensation, consciousness, and intuition propel us towards the skies.

How often do you have this kind of experience? Isn't this what it's all for? All the work. All the training. Conditioning. Technique. Graft. Craft. Trying. Striving. All that pressure, anguish and discomfort. Each ache and pain and injury. All the coaxing and cajoling. Crashing. Unpacking – processing – resetting. Every tear and ounce of sweat. Every failure. Every fall. Every joule taken to rise again, reach higher still, higher still, higher still… Every mite of will to push past every bloody 'I can't'. Every millisecond of practice and process is surely for this – to be limitless: to fly free.

Fluency is Freedom

Fluency is a practised ability; it releases us. In *technical* training, the rules and discipline of becoming experts in a particular code of moving are not an end in themselves. Mastery of technique, of 'speaking' that language fluently, allows you to communicate with nuance, wit, self-referentiality, ease, and speed. It helps you connect previously unconnected ideas or hijack existing knowledge, which can then be applied in any context and for any purpose. Technical fluency is a gateway to an ever-evolving practice, not a set of principles to attain and repeat as formulae. Fluency is action; physical intelligence is applied, not demonstrated.

Innovation arises from the effortless use of this knowledge; a dazzling tabula rasa of possibilities unfolds as the landscape of opportunity reveals itself.

Technical fluency is being present with the ebb and flow of your ageing body and knowing how best to calibrate, re-calibrate or reset its parameters for 'optimal' performance now: today, this hour, this minute, this second. It is 'in-body' intelligence directing and channelling perceptual and cognitive energies into an integrated whole. Technical fluency comes with body-positive confidence, a personal manifesto for trial and error, and a fearlessness to go to your outer limits. And, yes, sometimes that goes wrong – a misfire, you fall, fail, and break. More often, it gives you air to soar.

I think about the moments when I feel most 'in-body', most physically fluent. There are places one might expect it – on the stage or in an improvisation. These are situations where expert fluency is conjured on demand and shared – there is ego here, a pride in the mastery, the achievement, the work made visible. To be watched and seen is often motivation enough for further investment. But there are also the (un)exceptional moments of technical fluency. The altered stride length to catch the departing bus, dodging the crowd, balancing the dog and the groceries, saving the falling crockery, learning to paddle-board, ride a horse, fence, trek, swim in the ocean. To be in the body here is to live to the fullest.

Creative Fluency

We do not devote as much time to creative fluency as we do to technical fluency. We focus on positions, curricula, combinations, facts, figures, and rote memorisation. It can seem that the primary objective of fluency rests solely on the foundation of technical training.

But we rarely practise or teach the techniques of creativity. Why? Fluency demands accuracy and speed of execution. However, any expertly calibrated decision also needs a form of expression. Creative fluency demands all this – hard skills matched with a gentler but no less vital interpretative mode. Creativity is, to me, what makes life worth living. It surprises us, challenges our norms, tests our ethics, and questions our purpose. To live here is to live well and be alive.

Creative fluency is also about discovering multiple ways of solving problems and being open and curious enough to try them. It cannot be siloed into 'moments when you are creative'. It is a currency of everyday adaptability. I hope I am as creatively fluent in the kitchen as I am in dancing. And I hope to be as creatively fluent in relationships as I am choreographically. Creative fluency can be invited into any situation requiring resolution, test or investigation. It can be voiced anywhere, at any time.

Collaborative Fluency

In dance, fluency is always dependent on others. Everybody else must be in the right state for you to be able to perform to your highest level. Each dancer is rarely in sync, but that's what we're always aiming for because those shows where there's a shared fluency are exhilarating.

Sometimes, when we're lucky, we find it in the studio. The whole room, the dancers, and I are in this cohesive 'operating' space. There is an energy and a rush from this collective flow – a current connecting and charging us individually. Imagine the seeds of the white dandelion attached to the shoot – the fragile collection of nature's dots held by the most delicate of threads – rooting us back to the stem. The dandelion sways and bends in the wind, holding firm, flexing to the north, east, south and west in ever more acrobatic forms – holding as one, feeling as one. And then, caught on a thermal updraft, the seeds are released, with propellers everywhere – often travelling more than a kilometre from their point of launch – and land on the ground, where they too root and live again, forming a meadow of one million dandelions.

Here, fluency is a state of optimal performance, shared (sub)consciousness, and total absorption. The psychologist Mihaly Csikszentmihalyi called this a 'flow' state. As the hours pass and we journey on, we lose ourselves, with no roadblocks, not knowing what's next – a free-flow state between action and awareness, sensation and intuition. The self vanishes, the inner critic too, and we leapfrog from idea to idea. We are so immersed and blind to others' reactions, but even if we were to look up and look to the next, we would see their complete enthralment too – held captive by the same source – enchanted. Fluency takes us to a state of enchantment – why wouldn't you work to get there?

What is the work? It is placing yourself in a state of preparedness physically, intellectually and emotionally. Saturating your senses with inputs and stimuli that may overwhelm, test, excite, and terrify you. Research, seek and share – inside and outside of whatever box you have made for yourself. Invite others in and reach out to learn. Feel so full that you know with one more input, you will burst. And then quiet – create space between all that and you, let everything settle, seep and make its way through your body, a gentle diffusion, unhurried osmosis, ink through water.

We Divine

I open a portfolio of drawings – Robert Rauschenberg's haunting 'Dante' series of transfer drawings, images whose viscerality and energy have haunted and fascinated me for years – and place it on the studio floor. I remind myself of the artworks' properties and their feel. I select, more like settle, on one. I stay with it for a while. Music (not *the* music) is playing. The images within the image are burned into my retinas, imprinting on consciousness. I feel ideas stirring in my body – a point of initiation, a shape – a physical direction. The ideas swirl around and around until, eventually, one takes hold. I start to explore physically. And the choreography arrives! Not as predetermined steps, but in conversation with myself: body, head, head, body, body, body, head, head. Return to the image, draft.

A dance phrase starts to emerge way before it is set, and as an act of improvisation, I share it. Or, in other words, I play. And today's dancer, Ed, intuits, hijacking motion, grasping the essential. His version of the essential. He is fast. We start with an idea, he gets the sense of it, and we move on: word-building, phrase-building. I do not attempt to remember the material (although, strangely, I can with prompts); Ed remembers. Slower than the making, but solidly grafting, stitching together – Ed finds the phrase.

The direction is resolutely forward. We do not go back. When that idea has wholly passed, fizzled out, I catch another by its tail, refuelling the subsequent investigation – the next inevitable. Sometimes, I search for a new input, reach for another of many embodied options, or return to the portfolio.

We never speak: a total non-verbal exchange as the music plays on. Thirty minutes pass – maybe more. Consumed by the new solo, Ed thwarts being overwhelmed by reviewing and taking time to catch the whole. When he has the phrase, he dances it full out, and then again – embedding these signals into the fibres of his body, the fabric of his mind. And so, we continue.

Finally, I feel the movement inspiration pause, the ideas... and an unbelievable urge to see what we have made. What has been caught, what remains, what was lost? Together, we exchange what we remember. Ed has most of the phrase. I add details and specific action memories: the particular. And these moments are memories of the work that I will always carry, easily recalling them years or decades later, whenever the phrase is back in front of me.

Yet now, as I watch, there is a new referent: Ed. The physical decisions he has made. His inflexion, emphasis and embodied sensibility. I contrast his embodied intention with the body image I hold of the phrase's making and the imbued consciousness of the ideas' origins – intention searching again for its host/host searching for its intention. And I begin to add to this phrase, embroidering, editing, extending, refining, playing. Both as a means of consolidating the past and working out what it has become – this version of a version of a first draft that we have divined together. Who gave what to whom? Forever unclear.

Always, both: WE.

What does fluency really mean? At what point of learning can we genuinely say that we have become fluent? Well, to some degree, each of us is already physically fluent. From conception to death, we are constantly shifting body knowledge, learning and unlearning, adapting and applying. But mostly, we do this unconsciously, accidentally and 'naturally'.

I view fluency as a dialogue between what I know and what I seek. It involves connecting with rebellion, linking to strain, and transforming the statutes into new versions of ourselves.

But to get there, you need to have acquired a breadth of knowledge, a broad range of techniques, and a depth of experience from which to draw. The more you expand, accommodate, and imbue, the more access you have to misbehave more beautifully and more often. Real fluency allows you to operate in that language that sits outside the most common situations, out of the ordinary, to – dare I say it – occupy original space.

When teaching choreography, I share these tools with the dancers, and we practice them. But to attain real fluency, those artists must invent their own tools. They need to go to the next level to develop their fluency – it's not enough to rehearse the tools we've worked on. They must find their connections, their own ways in and out.

This book is an invitation to fluency. In it you've been given lots of ideas and principles, lots of tools. How you weld them together is personal; you just need the confidence to do it.

We Are Movement

By applying your burgeoning physical intelligence to your own life while progressing through these pages, you've moved closer to physical fluency. By working with and through your body and mind, you've begun to embed the skills, tools and techniques that form physical fluency's foundations. The groundwork's been done.

Think about it. You're now connected to your embodied mind and physicality, and perceptively attuned to your 360-degree surrounds. You're in command of your attention as it travels inside and outside of your body. You're alive to your momentary fluctuations of affect and energy. You're authentically present. Able to prepare yourself for any style of social interaction. You have an intense read on your physical signature and your expressive biases. And you can read other humans individually and through shifts in group dynamics. You can even read a packed room. You're more emotionally intelligent and kinaesthetically empathetic than you've ever been. You've reset your relationship with fear, manifested a fear-free space where you and your collaborators may freely take creative risks. And you are energised to reach new creative heights! You understand that your physical thinking is interactive and can be distributed between yourself and your collaborators. You've discovered you can visualise using any sensory system at all, including the somatic. Through these pathways, you can also access your original ideas, inspirations, and intuitions. You're attentionally primed. You've even (re)learned how to improvise!

You have the means to fly! All you need to do now is to keep practising these skills. Until all those that pertain to your own embodied life, creative practice and desired aims are available to you in a flash. Until you can flow between them seamlessly, because your instrument is as unique as any of the elite dancers I work with.

Close your eyes. Envision lights blinking on a near horizon. This is our physical thinking utopia. It's really that close. All you have to do is keep travelling towards it. You already are. Just take the next small step, and the next, and the next.

Inside us, we carry a tumultuous choreography of blood and fluid, organs, and joints. Our bodies hold floating ribs, mobile kneecaps and slipping discs; our tendons and ligaments strain and stress and strengthen and decay. We exist as a great organism of impermanence: our skin sheds, our hair falls out, our nails grow. We are attached to the revolving earth by gravity like pins in a pincushion, turning and turning in the cosmos. So many moving parts are waiting to be coordinated, made useful, and inspired. We embody what our body knows. We are creativity. We are interactive. We are communication. We are sensation. We are awareness.

WE ARE MOVEMENT

Acknowledgements

Book writing, as it turns out, is like dance making, a highly collaborative endeavour. This collective act of thinking together, inspiring one another, and critically challenging, testing, and interrogating ideas has been a fundamental pleasure in this herculean process of exchange. My interlocutors have been insightful, gracious, generous, sometimes argumentative, and always supportive – holding me in a space of the unfamiliar and encouraging me beyond my limits.

To Suze Olbrich, thank you for the hours, days, weeks, and months of conversation, interviews, research, words, dedication and understandings – your willingness to delve deeply into this complex world with me has been illuminating, thrilling, and formative. We now have enough material for multiple books and a connection beyond words.

Josh Ireland, thank you for your clarity and precision, as well as your openness and curiosity. You helped shine a torch on the essential.

Sir Ben Okri, thank you for that walk in the park and the advice on *not* staring at the blank page – how do you begin? To Ken Hollins, for your brilliant writing manifesto that I cradled next to my manuscript and often hijacked – a writing lifesaver.

And dearest Antoine Vereecken, thank you for your ears (and genuine positivity) as I read out my evolving chapters repeatedly during holidays, weekends away, and in airport lounges. Never once did you wince when you saw me pack that script – well, perhaps once! You truly deserve all the honours coming your way for being the most supportive audience ever, and remember, you added some beautiful and poetic interventions along the way.

Chris Wellbelove – this was all your idea. To invite me to consider writing a book and then to guide me so diligently, patiently, and with such care – you exceptional human. I am grateful for your kindness, flexibility, and astounding tenacity. It has been a phenomenal experience collaborating with you.

Alexis Kirschbaum and the entire Bloomsbury team – from the moment we met on Zoom during the pandemic to every stage of this process - you have been an exemplary facilitator. Gently creative, pushing when necessary, broadening the dialogue when appropriate, and making the overwhelming task of writing your first book feel somehow inevitable – effortless even. I realise that this conceals the intense work behind the scenes and the considerable effort and imagination required to bring a book into being. Like an iceberg with only its tip visible above the water, thank you for persuading me that I only needed to write the words.

To the entire Studio Wayne McGregor and Royal Ballet families – dancers, artists, collaborators, scientists, technologists, researchers, archivists, administrators, publicists, freelancers, mentors, friends, champions and donors, and to our extended colleagues across the world - thank you for believing and investing in the power of our physical intelligence. Sharing in the knowledge that the evolution and development of choreographic practice are essential attributes of living a full and intentional life fills me with joy. Knowing that you are with us on this journey of research and experimentation is fortifying – we are stronger together.

And finally, to all the dancers I have met and will meet, worked with and will work with, watched and will watch – if I have learned anything valuable in over thirty years of advocating for and loving dance, I have done so through and with you. Your special and unique knowledge, your insight, your capacity to sense, see and communicate, create, innovate and invent have always been my North Star. Within you, I find my purpose and focus. You are the light. The most heartfelt gratitude.

Notes

Many excellent sources have contributed to this book. I haven't included the full notes here, but those interested can find them at www.waynemcgregor.com/wearemovement

1: WE ARE ATTENTION

1 Proske, Uwe & Gandevia, Simon C., 'The Proprioceptive Senses: Their Roles in Signaling Body Shape, Body Position and Movement, and Muscle Force', *Physiological Reviews* 92(4), October 2012, pp. 1651–97. DOI: 10.1152/physrev.00048.2011

3: WE ARE 3D

1 Jola, C., Davis, A. & Haggard, P., 'Proprioceptive integration and body representation: insights into dancers' expertise', *Experimental Brain Research* 213, June 2011, pp. 257–65. DOI: https://doi.org/10.1007/s00221-011-2743-7; Karni, Avi & Meyer, Gundela & ReyHipolito, C.H. & Jezzard, Peter & Adams, Michelle & Turner, Robert & Ungerleider, Leslie, 'The acquisition of skilled motor performance: Fast and slow experience-driven changes in primary motor cortex', *Proceedings of the National Academy of Sciences of the United States of America* 95(3), March 1998, pp. 861–8. DOI: 10.1073/pnas.95.3.861; Morasso,

Pietro & Casadio, Maura & Mohan, Vishwanathan & Rea, Francesco & Zenzeri, Jacopo, 'Revisiting the Body-Schema Concept in the Context of Whole-Body Postural-Focal Dynamics', *Frontiers in Human Neuroscience* 9, February 2015. DOI: 10.3389/fnhum.2015.00083.

7: WE ARE TOUCH

1. David J. Linden, *Touch*, p. 64, Penguin Books (New York, 2016)
2. David J. Linden, *Touch*, p. 31, Penguin Books (New York, 2016); Hertenstein, M.J., Keltner, D., 'Gender and the Communication of Emotion Via Touch', *Sex Roles* 64, January 2011, pp. 70–80. DOI: https://doi.org/10.1007/s11199-010-9842-y
3. Field T., 'Massage therapy research review,' *Complementary Therapies in Clinical Practice* 24, April 2023, pp. 19-31. DOI: 10.1016/j.ctcp.2016.04.005
4. Kirsch, Louise & von Mohr, Mariana & Koukoutsakis, Athanasios & Fotopoulou, Aikaterini, 'Vicarious touch: a potential substitute for social touch during touch deprivation,' April 2024. DOI: 10.31234/osf.io/637xk.

10: WE ARE STORIES

1. Calbi, M., Angelini, M., Gallese, V. et al, ' "Embodied Body Language": an electrical neuroimaging study with emotional faces and bodies', *Scientific Reports* 7, 6875, July 2017. DOI: https://doi.org/10.1038/s41598-017-07262-0; Watson, R. & de Gelder, B., 'The representation and plasticity of body emotion expression', *Psychological Research* 84, January 2019, pp. 1400–1406. DOI: https://doi.org/10.1007/s00426-018-1133-1; de Gelder B., 'Why bodies? Twelve reasons for including bodily expressions in affective

NOTES

neuroscience', *Philosophical Transactions of the Royal Society of London. Series B, Biological Sciences* 364(1535), December 2009, pp. 3475–3484. DOI: https://doi.org/10.1098/rstb.2009.0190

11: WE ARE MEMORY

1   David J. Linden, *Touch*, p. 147, Penguin Books (New York, 2016).

12: WE ARE EMPATHY

1   Simonelli, F., Handjaras, G., Benuzzi, F., Bernardi, G., Leo, A., Duzzi, D., Cecchetti, L., Nichelli, P. F., Porro, C. A., Pietrini, P., Ricciardi, E. & Lui, F., 'Sensitivity and specificity of the action observation network to kinematics, target object, and gesture meaning', *Human Brain Mapping* 45(11), e26762. DOI: https://doi.org/10.1002/hbm.26762.

Bibliography

Armony, Jorge. *The Cambridge Handbook of Human Affective Neuroscience*. Cambridge University Press (2025).
Bailenson, Jeremy. *Experience on Demand*. W.W. Norton & Company (2018).
Bainbridge-Cohen, Bonnie. *Sensing Feeling and Action*. North Atlantic Books (1994).
Berger, John. *Ways of Seeing*. Penguin Books (1972).
Blakemore, Colin. *Images and Understanding*. Cambridge University Press (1990).
Biss, Eula. *On Immunity*. Fitzcarraldo Editions (2015).
Bourdieu, Pierre. *Outline of a Theory of Practice*. Cambridge University Press (1977).
Bown, Alfie. *Dream Lover*. Pluto Press (2022).
Bridle, James. *Ways of Being*. Allen Lane (2022).
Burton, Tara Isabella. *Strange Rites*. PublicAffairs (2020).
Cage, John. *Silence*. Wesleyan University Press (1961).
Claxton, Guy. *Intelligence in the Flesh*. Yale University Press London (2015).
Coates, John. *The Hour Between Dog and Wolf*. Fourth Estate (2012).
Coates, Ta-Nehisi. *Between the World and Me*. Text Publishing (2015).
Cunningham, Merce. *Changes: Notes on Choreography*. The Song Cave (2019).
Dale, Claire and Peyton, Patricia. *Physical Intelligence*. Simon & Schuster (2019).
Damasio, Antonio. *Descartes' Error*. Vintage (2006).

Damasio, Antonio. *Self Comes to Mind*. Vintage (2012).
Damasio, Antonio. *The Feeling of What Happens*. Vintage (2000).
Dixon, Steve. *Digital Performance*. Massachusetts Institute of Technology Press (2007).
Doidge, Norman. *The Brain That Changes Itself*. Penguin Books (2008).
Eagleman, David. *The Runaway Species*. Canongate (2017).
Eberhardt, Jennifer. *Biased*. Penguin Books (2020).
Ehrenreich, Barbara. *Dancing in the Streets*. Granta (2008).
Feldenkrais, Moshe. *The Potent Self*. North Atlantic Books (2003).
Feldman Barrett, Lisa. *How Emotions Are Made*. Macmillan (2017).
Foster, Susan. *Choreographing Empathy*. Routledge (2010).
Frantzis, Bruce. *Opening the Energy Gates of Your Body*. North Atlantic Books (2005).
Frayn, Michael. *The Human Touch*. Faber & Faber (2007).
Gallagher, Shaun. *How the Body Shapes the Mind*. Oxford University Press (2005).
Gardner, Howard. *Frames of Mind*. Basic Books (1993).
Gladwell, Malcolm. *What the Dog Saw*. Penguin Books (2010).
Goleman, Daniel. *Emotional Intelligence*. Bloomsbury Publishing (2004).
Gopnik, Alison. *The Gardener and the Carpenter*. Vintage (2017).
Gottlieb, Robert. *Reading Dance*. Pantheon Books (2008).
Grafton, Scott. *Physical Intelligence*. John Murray (2021).
Hartley, Linda. *Wisdom of the Body Moving*. North Atlantic Books (1995).
Hendren, Sara. *What Can a Body Do?* Riverhead Books (2020).
Hildyard, Daisy. *The Second Body*. Fitzcarraldo Editions (2017).
Hoffman, Eva. *Time*. Profile Books (2011).
Humphrey, Doris. *The Art of Making Dances*. Dance Books (1997).
Huvstedt, Siri. *The Delusions of Certainty*. Sceptre (2017).
James, William. *The Principles of Psychology*. Vol. 1 and 2. Dover Publications Inc (2000).
Kahneman, Daniel. *Thinking Fast and Slow*. Penguin Books (2012).
Kandinsky, Wassily. *Point and Line to Plane*. Dover Publications (2012).
Katan-Schmid, Einav. *Embodied Philosophy in Dance*. Palgrave Macmillan (2016).

Kurzweil, Ray. *The Singularity is Nearer*. Bodley Head (2024).
Laban, Rudolf. *The Mastery of Movement*. Dance Books (2011).
Leach, James. *Creative Land*. Berghahn Books (2004).
Levitin, Daniel. *This is Your Brain on Music*. Atlantic Books (2008).
Linden, David J. *Touch*. Penguin Books (2016).
Lobel, Thalma. *Sensation*. Icon Books (2014).
Lorde, Audre. *Your Silence Will Not Protect You*. Silver Press (2017).
Marshall, Lorna. *The Body Speaks*. Methuen Drama (2008).
Martin, Betty. *The Art of Receiving and Giving*. Luminaire Press (2021).
May, Rollo. *The Courage to Create*. W. W. Norton & Company (1994).
McCully-Brown, Molly. *Places I've Taken My Body*. Faber & Faber (2021).
Merleau-Ponty, Maurice. *The Phenomenology of Perception*. Routledge (2013).
Mukherjee, Siddhartha. *The Gene*. Vintage (2017).
Naisbitt, John. *High Tech High Touch*. Nicholas Brealey Publishing (2001).
Newen, Albert. *The Oxford Handbook of 4E Cognition*. Oxford University Press (2018).
Newlove, Jean. *Laban for Actors and Dancers*. Nick Hearn Books (1993).
Noë, Alva. *Action in Perception*. Massachusetts Institute of Technology Press (2006).
Novack, Christine. *Sharing the Dance*. University of Wisconsin Press (1990).
Odell, Jenny. *How to Do Nothing*. Melville House Publishing (2019).
Pang, Camilla. *Explaining Humans*. Viking (2020).
Paxton, Steve. *Gravity*. Contredanse (2018).
Piaget, Jean. *The Psychology of Intelligence*. Routledge (2001).
Pineda, Jaime. *Mirror Neuron Systems*. Humana Press (2010).
Ridley, Matt. *Nature via Nurture*. HarperPerennial (2004).
Robinson, Ken. *Finding Your Element*. Penguin Books (2014).
Rodenburg, Patsy. *Presence*. Penguin Books (2009).
Rovelli, Carlo. *The Order of Time*. Penguin Books (2019).
Rubin, Peter. *Future Presence*. HarperOne (2020).

Sacks, Oliver. *The Man Who Mistook His Wife for a Hat*. Picador (2011).
Spence, Charles. *Gastrophysics*. Viking (2017).
Todd, Mabel E. *The Thinking Body*. Princeton Book Company (1980).
van der Kolk, Bessel. *The Body Keeps the Score*. Penguin Books (2015).
Watson, Gay. *Attention*. Reaktion Books (2017).
Williams, Mark and Wigmore, Tim. *The Best*. John Murray Business (2020).
Woolf, Virginia. *Moments of Being*. Vintage (2002).

Index

ABBA *Voyage*, 178–83
Acosta, Carlos, 138, 142–3
Action Observation Network, 224
acupuncture, 116
adaptors, 155–6, 210
adrenaline, 86–7, 107, 149, 207, 234
aerialists, 68
aikido, 64
Albers, Josef, 255
allostasis, 99
Amazon, 109
Anderson, Paul Thomas, 174
anxiety, 60–2, 84, 93, 97, 108, 116, 123, 155–6, 175, 197–8, 243, 245–7
  anti-anxiety drugs, 199
  performance anxiety, 206–10, 250
  social anxiety, 126
Apple Health app, 109
archery, 69
Aristotle, 117
arousal levels, 86–7, 90–4, 97, 100–1, 108, 117, 140, 213, 217, 259
Artist, The, 174
Astaire, Fred, 153
AtaXia, 15
Atomos, 87–8

attention, 5–19
  and proprioception, 12–18
  qualitative vs. quantitative, 8–9
  audience–performer bio-loops, 149
authoritarianism, 109
avatars, 41, 176, 178–83
Ayurvedic kitchen, 102

Bach, J. S., 255
back injuries, 44–5
backache, 82
backspace, 6, 43, 52
Bacon, Francis, 255
balance, 4, 12, 16, 50, 53, 56–9, 61–6, 118, 130, 208, 221, 229–30, 249
ball pools, 63
ballet, 27, 37, 42, 119, 147–8, 216, 229, 240, 242–3
  and gesture, 158–62
  see also Royal Ballet
ballet bars, 185
Ballet Frankfurt, 216
barefoot walking, 129–30
Barnard, Phil, 262
Barrett, Lisa Feldman, 101
battements, 159
Beau Travail, 174
Belfast, 122

Bharatanatyam dance, 37, 159
Biden, Joe, 157
Biles, Simone, 106
biological motion, human ability to identify, 182
biometric technologies, 87–8, 109–10
blind lead, 123–6, 128, 131–2
blood pressure, 93, 101, 110
blood sugar levels, 91, 102
body clocks, 20
body image, 47, 71, 93, 107–8, 285
body language, 175–8, 186, 188, 215, 226
  and social interactions, 186–9
  and territorial behaviour, 183–5
body maps, 118–19
body-popping, 26, 79
body schemas, 46–9, 52–4, 103, 117, 136, 275
  see also movement schemas
body temperature, 95, 116
Bosnia and Herzegovina, 122
brain–body disorders, 17
breathing, 6, 10–11, 38–9, 57–8, 84–6, 91–6, 109, 116, 124, 152, 173–5, 228, 250
  and anxiety, 197, 199–200, 206–9
  and audience feedback, 148–9
  and avatars, 182–3
  Breath of Fire technique, 95–6
  breathwork exercises, 8, 95, 206, 212
  circular breathing, 93–4
  deep breathing, 67, 82, 92–3, 95, 213, 246
  and heart rate, 199–200
  holding breath, 166, 169, 199
  rapid breathing, 101
  and social interactions, 193–4

British Council, 122
BWM, 22
Byrne, Alexandra, 166

C-tactile system, 202–3, 205
calorie counting, 7, 19, 108
can-can, 75
capoeira, 36, 75
card games, 178
cardiovascular system, 82, 101, 209, 249
Caspersen, Dana, 216
central nervous system, 56
chaining, 79–81
Chaplin, Charlie, 174
charades, 178
choreographic process, 247, 270–1, 273–4, 291
chronic pain, 11, 196, 205
City Lights, 174
Cleansed, 163
Company Wayne McGregor, 15, 28, 76, 162, 251, 256, 272
consciousness, 99, 117
Contact Improvisation, 119, 125–6, 128, 132–3
coordination, 4, 21, 27, 50, 52, 58, 69–83, 105, 124–5, 201, 218, 249, 256, 275, 288
  complex coordination, 73–81
  and motor skill development, 71–3
  and play, 70–1
  and proprioception, 71–2
coronavirus pandemic, 116, 126–7, 245–6
cortisol, 82, 86, 107, 130, 234, 246
cranial osteopathy, 103
CrossFit, 7
Csikszentmihalyi, Mihaly, 282

INDEX

Daley, Tom, 80–1
Dance Alloy, 3
dance analysis, 190–3
Dancing at Lughnasa, 164
Day-Lewis, Daniel, 174–5
de la Tour, Frances, 163
deep-sea diving, 26
deLahunta Scott, 262
dental work, 199–201, 203–4
depression, 40, 91, 100, 140, 140
développes, 210
distributed cognition, 272
divergent thinking, 233, 240–4
diving, 80–1
dopamine, 107–8, 207, 222
dyspraxia, 17

Ekman, Paul, 155
emblems, 157–8
empathy, 17, 97, 168, 175, 214–30
   kinaesthetic empathy, 136, 194, 217–30, 272, 275
   and mirror neurons, 223–4
   mirroring and matching, 186–7, 226–9
   and nonverbal cues, 214–16, 218
endocrine system, 79
endorphins, 101
energetic baseline, 89–91, 164, 197
English Patient, The, 168
ensōs, 278
enteric nervous system, 100
equilibrium point (centre of gravity), 62–3
ergonomics, 30, 33
Event Horizon, 11
exhaustion, 102

eye contact, 123, 136, 149, 152, 175, 178, 182, 189, 192–3, 216

Facebook, 108
facial expressions, 136, 174–7, 179–80, 182, 186, 188, 192, 219
facial sensitivity, 118–19
falling, 56–64, 66–8, 162
Fame, 269
Fantastic Beasts, 163
Fargo, 168
fasting, 105
fear, need for, 234–6
feedback, 5–6, 10, 118, 129, 149, 172, 221, 225–6
feet, and balance, 64–6
fencing, 69, 217
fidgeting, 155–6, 210
fight-or-flight, 94, 234
film scores, 168
first dates, 84–5, 94
Fitbits, 7, 87, 109
flat dancing, 42, 49
flow states, 139, 282
fouettes, 229–30
free play, 258
free radicals, 130
free-running, 63, 68
Friesen, Wallace V., 155

gait, 20–1, 35, 41, 129
gender norms, 23–5
gestural language, 22–3, 153–5, 157–62, 166–7, 169–72, 178, 215
Gorman, Amanda, 157–8
Gormley, Antony, 11
Graham, Martha, 26
grief, 140
grooming, 115, 155
gymnastics, 68, 106, 120, 217, 221

handedness, 30
hands, brain perception of, 155
Harry Potter, 163
hearing, 12, 18, 39, 96, 235
'heart brain', 100
heart rate, 7, 87, 95, 101, 103, 116, 199–200, 204, 259
heartbeat, 57–8, 93
hip-hop, 36, 75
Hof, Wim, 93
homeostasis, 99
homophobia, 23, 109
homunculus, 118
horse riding, 63, 138, 217
humour, and performance anxiety, 209
hunger, 102

ice skating, 63
illustrators, 156–7
improvisation, 17, 27–8, 36, 167, 169, 195, 206, 215, 225, 234, 243–4, 246, 250, 255–7, 274, 280, 284, 287
see also Contact Improvisation
Industrial Light & Magic, 180
inflammation, 100
Infra, 162
Innsbruck Goggle Experiments, 15–17
insulin, 7
interactive cognition, 275
interoception, 98–103
online, 107–10
intimacy, 116, 121–2, 126, 136, 141, 162, 214–16, 220, 229–30

jazz, 37
Jennings, Garth, 225
jetés, 207

Joan of Arc, 174
Joker, The, 168
Judson Church, 161
juice detox, 102

Kane, Sarah, 163
Keaton, Buster, 174
kinaesthetic images, 97
kinespheres, 50, 52–5, 121, 164, 189, 254
Kirsh, David, 272

La Scala, 66
Laban, Rudolf, 49–50
Lego, 258
Leonardo da Vinci, Vitruvian Man, 41, 50
lifestyle, 6, 30–3
lifestyle apps, 107
Limón, 37
Lloyd, Harold, 174
London Underground (tube), 38–40, 46, 85

marathon runners, 91
marking, 158–9
'marking for self', 276
martial arts, 64
Mary Queen of Scots, 164–70
masculinity, 24
matching, *see* mirroring
meditation, 82
Mehling, Wolf, 99
memory, 20, 40, 76–7, 97, 101, 134, 159, 162, 190, 195–6, 198, 200, 208, 272–3
menstruation, 100
mental health, 82, 198
mental imagery, and creativity, 255–68

INDEX

#MeToo Movement, 119
Metropolis, 174
microaggressions, 215
Miller, Ezra, 163
Mind and Movement technique, 263–8
mindfulness, 82
mirror cells, 194, 223–4
mirroring (and matching), 71, 186–7, 191–2, 215, 225–9
mirrors, 40–2, 192
misogyny, 184
moods, 102–3
movement direction, 162–5
movement schemas, 45–8, 71, 196, 218–19, 223, 233
  see also body schemas
mudras, 159
muscle memory, 28
musculoskeletal degeneration, 11
musicality, 18, 57

Nadal, Rafael, 211–12
National Theatre, 164
negativity bias, 245
No One Is an Island, 22

Obama, Barack, 144
osteopathy, 199
Oura, 109
oxygen, 7, 86, 92, 94
oxytocin, 86, 107, 135, 205

pain perception, 201–6
paleo diet, 102
paragliding, 26
parasympathetic nervous system, 93
parsing, 74–6, 78, 131–3, 135
peri-personal space, 121–3, 164, 189

phobias, 235
physical signatures, 7, 21–31, 37, 72, 138, 156, 163, 167, 177, 179, 228, 248, 286
piano playing, 19, 69, 83, 275
pirouettes, 33, 159, 207, 223
Pittsburgh Dance Council, 3
pliés, 8, 207
posture, 27, 30, 62–3, 151, 175, 179, 188, 191–2, 226–7, 262, 265
  see also body language
prayer, 211–12
proprioception, 12–18, 21, 46–7, 103, 202, 254, 259
  and balance, 56, 58–9, 63–4
  and coordination, 71–2
  and reflexes, 67–8
  and touch, 117–18
PTSD, 126
public speaking, 84–5, 94

racism, 109
Radiohead, 225
Random International, 22
Rauschenberg, Robert, 283
reflexes, 67–8
reflexology, 83
reiki, 199
repeat stretches, 207
rituals, 211–13
road safety, 53–4
Robbie, Margot, 165, 167–8
rock climbing, 26, 63
Rodenburg, Patsy, 139, 145, 152
Romania, 122
Ronan, Saoirse, 165, 167–70
Ronson, Mark, 255
Rourke, Josie, 164, 166
Royal Ballet, 89, 162, 251, 260, 272

Royal Institution, 84–5
Royal Opera House, 185
Rubber Hand Illusion, 136

Sadler's Wells, 216
salsa, 34
saunas, 102
self-awareness, 111, 152, 171
self-care, 104–7
self-conception, 47, 248
self-confidence, 25, 212
self-consciousness, 243
self-knowledge, and fear, 236
self-massage, 238
self-recording, 170–2
self-regulation, 210
self-stroking, 205–6
self-touch, 134–5
self-worth, 108
Senegal, 122
Serbia, 122
Seymour, Lynn, 153
Shakespeare, William, 139
shiatsu, 102, 116, 198–9
Siberia, 122
situational (three-dimensional) awareness, 40–5, 49–55, 94
Skarsgård, Alexander, 145–6
skateboarding, 63
skin sensitivity, 127–8
sleep, 20, 30, 32, 57, 77, 82, 91, 98, 102–4
Sleeping Beauty, 229–30
smartphones, 109
smell, 12, 22, 39, 117
social dance, 116
social media, 108, 217, 245
somatic markers, 97, 196–7
somatosensory cortex, 118
spatial perception, 39–40, 164
speed skating, 217
sports massage, 116
step counting, 7
stone-carving, 69
stress reduction, 116
stress response, 79, 82, 86
Studio Wayne McGregor, 22, 42, 122, 185, 247, 260, 263
Studio XO, 87–8, 110
sugar fasts, 102
surfing, 63, 221
swimming, 103

taste, 12, 117
tattoos, 176, 201, 203–4
technical fluency, 279–81
tennis, 138, 211
territorial behaviour, 183–5
theatre bars, 185–6
therapeutic dance, 237–9
There Will Be Blood, 174
'today body', 9
touch, 12, 22, 115–37
  and body maps, 118–19
  dual nature, *see* C-tactile system
  and parsing, 131–3, 135
  self-touch, 134–5
  touch deficit, 127
  vicarious touch, 136
touch-typing, 69, 83
transphobia, 109
Turkey, 122

'uncanny valley' concept, 182
understudies, 272

veganism, 102
vestibular system, 56, 118

INDEX

virtuosity, 8, 19, 28, 69, 72, 83, 216, 221, 230
vision, 12–18, 56, 71, 96, 117, 235

West Coast swing, 26
whitewater rafting, 35

xenophobia, 109

yoga, 7–8, 34, 64, 95, 129
York, Thom, 224–6

Zen Buddhism, 278
Zorbing balls, 50

A Note on the Author

Professor Sir Wayne McGregor CBE is a multi-award-winning British choreographer, director and curator pursuing physical intelligence through body and technology. He is the founder of Studio Wayne McGregor in London, a studio encompassing creative collaborations in dance, choreography, film, music, visual art, technology and science; a company of dancers, Company Wayne McGregor; and learning and research programmes. McGregor is Resident Choreographer at the Royal Ballet and Director of Dance for La Biennale di Venezia. He is also the choreographer of *ABBA Voyage*, the revolutionary avatar concert. His work continues to be performed and exhibited internationally.

Professor of Choreography at Trinity Laban Conservatoire of Music and Dance, McGregor's research residencies include: Innovator in Residence at UC San Diego and Research Fellowships at King's College, Cambridge and Oxford University's Schwarzman Centre of Humanities. McGregor has Honorary Doctorates from The Royal College of Art London, Plymouth University, University of Leeds, University of Chester, and UAL, and is an Honorary Fellow of the British Science Association.

A Note on the Type

The text of this book is set in Bembo, which was first used in 1495 by the Venetian printer Aldus Manutius for Cardinal Bembo's *De Aetna*. The original types were cut for Manutius by Francesco Griffo. Bembo was one of the types used by Claude Garamond (1480–1561) as a model for his Romain de l'Université, and so it was a forerunner of what became the standard European type for the following two centuries. Its modern form follows the original types and was designed for Monotype in 1929.

WE ARE MO>EMENT
WE ARE MOΛEMENT
WE ARE MO<EMENT
WE ARE MOVEMENT
WE ARE MO>EMENT
WE ARE MOΛEMENT
WE ARE MO<EMENT
WE ARE MOVEMENT
WE ARE MO>EMENT
WE ARE MOΛEMENT
WE ARE MO<EMENT
WE ARE MOVEMENT
WE ARE MO>EMENT
WE ARE MOΛEMENT
WE ARE MO<EMENT
WE ARE MOVEMENT